FIELD EFFECT ELECTROOSMOSIS - A NOVEL PHENOMENON IN ELECTROKINETICS AND ITS APPLICATIONS IN CAPILLARY ELECTROPHORESIS

Edited by **Kiumars Ghowsi**

Field Effect Electroosmosis - A Novel Phenomenon in Electrokinetics and its Applications in Capillary Electrophoresis

http://dx.doi.org/10.5772/58644
Edited by Kiumars Ghowsi

Contributors

Constantina Kapnissi-Christodoulou, Malgorzata Katarzyna Sulkowska, Justyna Nowakowska, Elsa Lamy, Lénia Rodrigues, Martine Morzel, Ana Rodrigues Costa, Celia Antunes, Fernando Capela-Silva, Cristina Pinheiro, Orlando Lopes, Ma. Elena Páez-Hernández, José A. Rodríguez-Ávila, Emanuel Martínez, Israel Ibarra, José Manuel Miranda, Carlos A. Galán-Vidal, Kiumars Ghowsi

Published by InTech

Janeza Trdine 9, 51000 Rijeka, Croatia

Publishing Process Manager Iva Lipovic

Technical Editor InTech DTP team

Cover InTech Design team

First published February, 2015

Printed in Croatia

Additional hard copies can be obtained from orders@intechopen.com

Field Effect Electroosmosis - A Novel Phenomenon in Electrokinetics and its Applications in Capillary Electrophoresis, Edited by Kiumars Ghowsi

p. cm.

ISBN 978-953-51-2025-4

For EU product safety concerns:
IN TECH d.o.o., Prolaz Marije Krucifikse Kozulić 3, 51000 Rijeka, Croatia,
info@intechopen.com or visit our website at intechopen.com.

Meet the editor

Kiumars Ghowsi obtained a B.S. in electrical engineering science in 1981, and a M.S. in electrical engineering science in 1984 from Louisiana State University, USA. Professor Ghowsi changed his field to chemistry and obtained a Ph.D with a major in analytical chemistry and a minor in physical chemistry from Louisiana State University, USA, in 1990. Currently Professor Ghowsi is a member of IEEE. He has been a faculty member at Texas Tech Chemistry. After moving to Iran, he became a faculty member in Chemistry Department, Majlesi Branch, Islamic Azad University, Iran. His research interest are home made capillary electrophoresis with contactless electrode as a detector, new electrokinetic phenomenon called field effect electroosmosis and a new look at capillary electrophoresis and micellar electrokinetic capillary chromatography.

Contents

Preface

Electrokinetic phenomena were discovered quite early in the nineteenth century. Hence investigations in the field have been conducted for more than a century and a half. The discovery of electroosmosis and electrophoresis by Reuss occurred soon after the first investigations on the electrolysis of water by Nicholson or Carlisle and the electrolysis of salt solutions by Berzelius (1804) and Davy (1807). Reuss carried out two experiments, the first demonstrated the effect known as electroosmosis, and the second was actually the discovery of electrophoresis.

Field effect electroosmosis is a novel interfacial phenomenon which is of particular interest. Field effect could be demonstrated by combining a metal-insulator-electrolyte system (MIE) with capillary electroosmosis. Using this technique a capillary at its outside surface and electroosmatic flow is controlled by applying a perpendicular electric field to the flow.

The potential application of this effect is examined in the process which would benefit from a flexible control of electroosmatic flow, such as capillary electrophoresis in separation science.

Electrophoresis experiment was first carried out by Tiselius in 1930. In his thesis titled "The Moving Boundary Method of Studying the Electrophoresis of Proteins", Tiselius utilized the electric charge carried by the macromolecules to achieve some pioneering separation of blood plasma proteins in free solution on a photographic film. Application of electrophoresis experiment is defined as the transport of electrically charged particles in a direct current electric field. Electrophoretic separation is based on differential rates of migration in the bulk of the liquid phase and is not concerned with reactions occurring at the electrodes. In the early days, electrophoresis was carried out either in free solution or in the supporting media such as paper, cellulose acetate, starch, agarose, and polyacrylamide gel. Between 1950 and 1970, an enumeration of techniques and instrumentation for electrophoresis were developed.

Gel electrophoresis has been rarely used for the separation and identification of small charged molecules of molecular weight less than about 1000 Dalton. In addition, the major draw back of gel electrophoresis is lack of complete automation.

To overcome the low efficiency and reduce thermal effects, Hjerten carried out electrophoresis in narrow diameter tubes of 300 μm internal diameter for the first time in 1967.

This was the birth of open tubular capillary electrophoresis. However, in the following decade, capillary electrophoresis did not draw enough attention from researchers, until 1981 when Jorgenson and Lukacs demonstrated the use of narrow capillaries to produce high efficiency for the separation of dansy and fluorescamine derivatives of amino acids, dipeptides and simple amines, high performance capillary electrophoresis was born and a new era of capillary electrophoresis began. After the introduction of commercial CE instrument in late 1988 that allowed the full automation of CE analysis to be possible, more and more research publications and industrial applications have made capillary electrophoresis one of the dominant technologies in the separation field. In 1985 Terabe et al. added a new dimension to capillary electrophoresis. They added micelles to the aqueous electrolyte and were able to separate neutral molecules such as benzene and phenol.

With this technique it is possible to separate various drugs which are neutral or even charged. One that can separate enantiomers by this technique is called micellar electrokinetic capillary chromatography (MECC).

The present book contains few fundamentals on capillary electrophoresis and diverse applications of electrophoresis in general. We hope this collection will be useful for the interested readers.

Kiumars Ghowsi
Department of Chemistry,
Majlesi Branch,
Islamic Azad University,
Isfahan, Iran

Section 1

Field Effect Electrophoresis

Chapter 1

Field-Effect Electro-osmosis – a New Dimension in Capillary Zone Electrophoresis

Kiumars Ghowsi, Hosein Ghowsi and M. Razazie

Additional information is available at the end of the chapter

http://dx.doi.org/10.5772/59114

1. Introduction

Electrokinetic phenomena were discovered quite early in the 19th century. Investigations in the field have therefore been conducted for more than a century and a half. The discovery of electro-osmosis and electrophoresis by Reuss occurred soon after the first investigations on the electrolysis of water by Nicholson and Carlisle [1] and the electrolysis of salt solutions by Berzelius (1804) and Davy (1807). Reuss [2] carried out two experiments: the first demonstrated the effect known as electro-osmosis, and the second was the discovery of electrophoresis. Considering the simplest case of electro-osmosis in a single capillary, Helmholtz [3] obtained a formula for the linear velocity of electro-osmosis:

$$v_{eo} = -\frac{\varepsilon\xi}{4\pi\eta}E$$

where ξ is the interfacial electric potential difference, E is the electric field strength, and η is the viscosity of the liquid. Field-effect electro-osmosis is a novel interfacial phenomenon which is of particular interest. Field effect can be demonstrated by combining a metal-insulator electrolyte system (MIE) with capillary electro-osmosis. This technique uses a capillary at the outside surface, and electro-osmatic flow is controlled by applying a perpendicular electric field to the flow.

Potential applications of this effect could benefit from a flexible control of electro-osmatic flow, for example by capillary electrophoresis in separation science [4].

1.1. Theory

As a result of the elaboration of the theory, and in particular because of Saxen's experiments [5], entirely new premises appeared for experimental research. Not only did investigations of electrokinetic phenomena become possible at this stage, but studies of the double layer on the basis of these phenomena were also conducted.

The effect of the chemical nature of the surface and the ionic composition of electrolytes on the sign and magnitude of the potential (determined by the electrokinetic measurement) was determined experimentally in the early parts of this century, and furnished the grounds for solving the problem of the mechanism of the formation of the double layer of colloid particles.

Freundlich [6] called attention to a possible connection between the appearance of the double layer and an adsorption phenomenon.

It was considered that, if the adsorption coefficients of the ions were different, the strongly adsorbed ions would be present in excess on the surface, and the weakly adsorbed ions would be present in excess in the liquid part of the double layer, together giving rise to the double layer.

Another possible mechanism for the formation of the double layer is linked with the dissociation of surface ionogenic groups under the influence of a polar dispersion medium. This mechanism was first studied in regard to proteins. The ionogenic groups in proteins are of different chemical natures (acidic carboxyl, basic amino groups, etc.), and proteins are classified as amphoteric electrolytes.

In a first approximation, the amphoteric nature of monomer units of the protein molecule may be characterized by the following model:

$$R\begin{cases} NH_2 \\ COOH \end{cases} \begin{matrix} \rightarrow \\ \leftarrow \end{matrix} \quad R\begin{cases} NH_3^+ \\ COO^- \end{cases}$$

At low pH, the protein carries a + charge. As pH is increased, the isoelectric point is first reached, and then there is a change in the sign of the charge on the protein.

Chemical groups on the insulator surface, at an interface between a liquid and an insulating solid, dissociate similarly to the above mechanism. Due to this surface ionization and specific adsorption, the interface is charged and ions of opposite polarity to the interfacial charge (counter-ions) are attracted to it, while ions of the same polarity are repelled.

2. Theoretical modelling of the Metal-Insulator Electrolyte (MIE) and postulation of a novel electrokinetic effect called field-effect electro-osmosis

The zeta potential, which is the potential across the surface of the insulator and electrolyte, has been shown in the past to be manipulated by pH (surface ionization) and ionic concentration (specific adsorption). A novel phenomenon is postulated by us in which the zeta potential in a capillary can be controlled by an external field. If a thin-wall capillary is coated with a metallic conductor on the external surface and a voltage V_G is applied between the metal electrode and the electrolyte (Fig.1), electro-osmotic flow can be controlled. By changing V_G, the voltage drop across the double layer changes, and this change includes the change in zeta potential, ξ, which in turn causes a modification in the electro-osmotic flow. We name this phenomenon "external field-effect electro-osmosis," or simply " field-effect electro-osmosis". Therefore, zeta potential, ξ (pH,C, V_G), is a function of three variables: pH, C (the ionic concentration in the electrolyte), and V_G. To understand this approach, there is a need to model the metal-insulator-electrolyte system in more detail. This is addressed in the following sections.

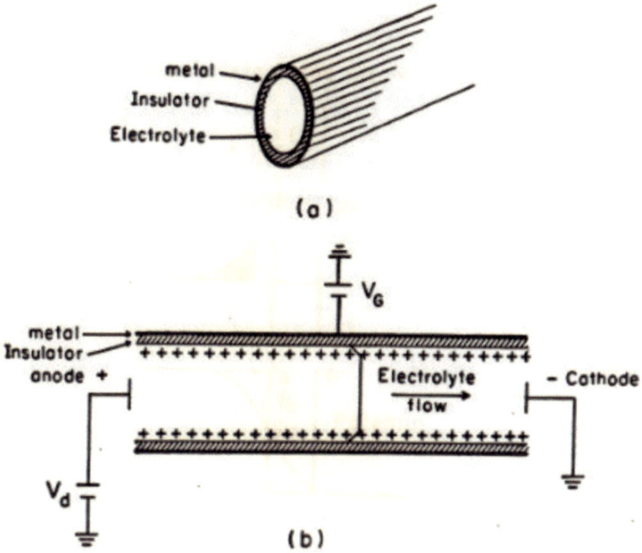

Figure 1. a) A Capillary covered with metallic coating; b) Cross-section of metal-insulator electrolyte, field-effect electro-osmosis

3. Ideal metal-insulator electrolyte structures

The ideal metal-insulator electrolyte (MIE) system is similar to what Siu et al. [7] have defined as the totally blocked interface of an insulator/electrolyte. In an ideal MIE, there is a complete absence of interfacial reactions between the electrolyte and oxide; in other words, there is neither specific adsorption nor surface ionization. Since the interfacial electrochemical processes are absent, the charge and potential distribution in this MIE system are dictated solely by electrostatic considerations. As shown in Fig. 2, the metal electrode is chosen to be ground and the applied voltage, V_G, is applied to the reference electrode. The reference electrode is chosen as nonpolarizable, where the voltage drop across it is negligible, so it can be assumed that V_G is applied to the electrolyte. The charge per unit area and the potential in the electrolyte space-charge region are related by the Poisson-Boltzmann equation [8]. From Gauss's law and the solution to this equation, we find that for an electrolyte the charge per unit area in the Gouy-Chapman space-charge region is given by:

$$\sigma_d = \left(8\varepsilon_e kTn_0\right)^{1/2} \sinh\left[q\left(V_G - \varphi_d\right)/2kT\right] \tag{1}$$

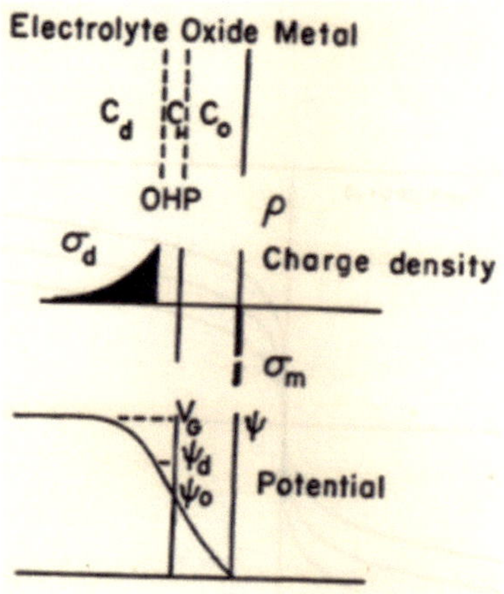

Figure 2. The charge and potential profile in a totally block metal-insulator electrolyte (MIE)

From Fig. 2, a charge neutrality equation can be written as follows:

$$\sigma_d + \sigma_m = 0 \tag{2}$$

Also, $C'_o = \dfrac{C_H C_o}{C_H + C'_o}$ since $C_H >> C_o$ and

$$C'_o \cong C_o \tag{3}$$

since $C_d >> C_o$ then $\varphi_d - \varphi_o << \varphi_o$, or

$$\varphi_d \approx \varphi_o \tag{4}$$

and

$$\varphi_o C_o \approx -\sigma_m \tag{5}$$

http://dx.doi.org/10.5772/59114

$$-\sigma_m = \sigma_d \approx C_o \varphi_d \tag{6}$$

Thus, from eqs. (1) to (6) we find that:

$$C_o \varphi_d = -(8\varepsilon_e kTn_0)^{1/2} \sinh[q(\varphi_d - V_G)/2kT] \tag{7}$$

Rearranging eq. (7), we obtain:

$$V_G = \frac{2kT}{q} \sinh^{-1}\left(\frac{\varphi_d C_o}{(8\varepsilon_e kTn_o)^{1/2}}\right) + \varphi_d \tag{8}$$

$$\xi \approx \frac{2kT}{q} \sinh^{-1}\left(\frac{\varphi_d C_o}{(8\varepsilon_e kTn_o)^{1/2}}\right) + \varphi_d \tag{9}$$

Since the difference between V_G and φ_d is very small (in the range of millivolts), whereas the difference between V_G and φ_d is several volts, by substituting φ_d into the parenthesis of eq. (9) V_G, the calculations are simplified. Thus, the following equation is found for the zeta potential:

$$\xi \approx \frac{2kT}{q} \sinh^{-1}\left(\frac{\varphi_d C_o}{(8\varepsilon_e kTn_o)^{1/2}}\right) \tag{10}$$

4. Contributions

It is interesting that much of the basic science involved in electrokinetic phenomena was discovered more than a century and a half ago[1]. After the discovery of dissociation of water by electricity and the scientific curiosity that ensued, electrokinetic phenomena were discovered in parallel with electrolysis of water. If a V-shaped test tube is filled with soil and with electrolyte, and a direct current voltage is applied across the soil, electrolyte is pumped from one side to the other side. This electrokinetic phenomenonis is called electro-osmosis. Electro-osmosis can occur at a capillary as well. The charge at the interface of the wall of the capillary is forced by the electric field applied across the capillary. We propose field-effect electro-osmosis, a novel phenomena where the zeta potential, ξ, is proportional to the charge at the interface of oxide and electrolyte, and V_G is the voltage perpendicular to the interface.

Fig. 3 shows two-dimensional zeta potential as a function of V_G at different concentrations of electrolytes. The basic science was electrokinetic.

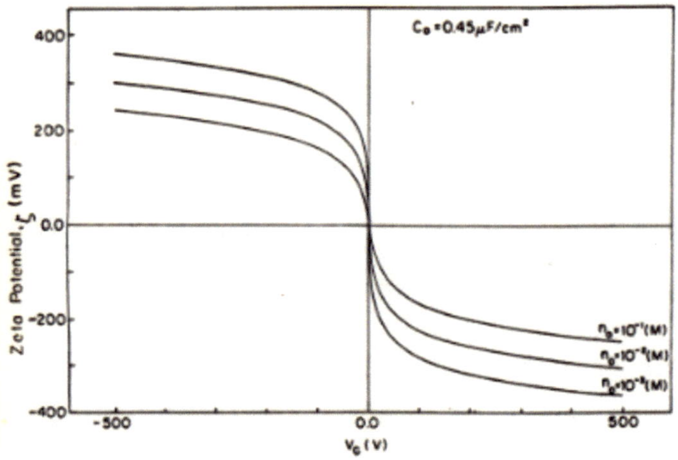

Figure 3. Change in zeta potential ξ as a function of V_G for various ionic concentrations for ideal case

Gradually during the past 30 years, CZE (capillary zone electrophoresis) [9] and micellar electrokinetic capillary chromatography (MECC) [10] have become applied sciences in their own right, through several publications [11]. Fig. 4 shows field-effect electro-osmosis at work in separation. The electro-osmosis in Fig.4 uses $V_d - V_G$ to make the voltage at the interface uniform. With field-effect electro-osmosis one can make the zeta potential zero or positive or negative.

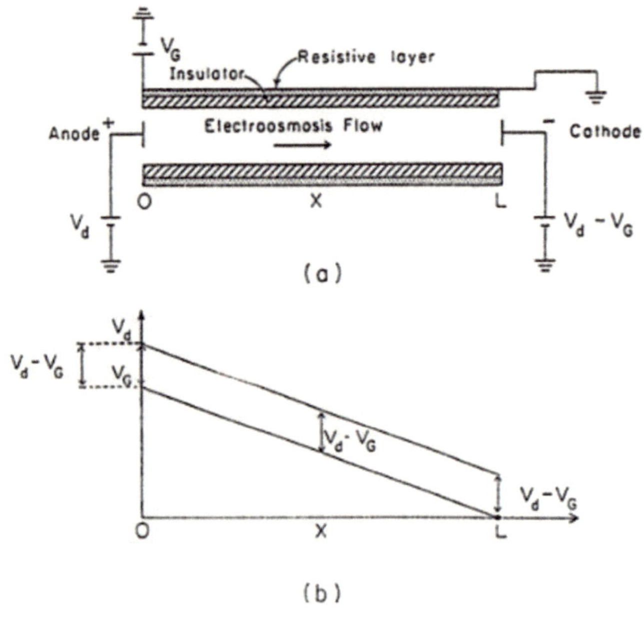

Figure 4. a) The schematic of field-effect electro-osmosis with the constant zeta potential across the capillary; b) The voltage perpendicular to the wall of the capillary versus X

This can achieve the separation of protein with less tailing or shift the movement of electro-osmosis to the left or right [12] (Fig. 5).

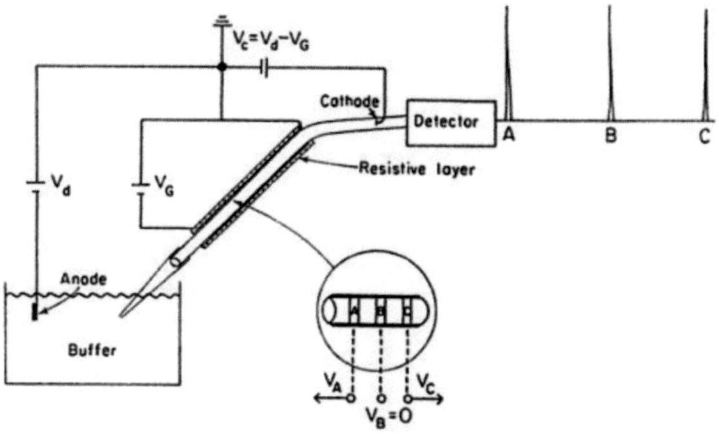

Figure 5. Field-effect electro-osmosis used in separation (source: Biosersor)

List of symbols

C_1 Inner Helmholtz layer capacitance

C_2 Outer Helmholtz layer capacitance

C_a Capacitance due to variation of surface charge σ_0 on the insulator with surface potential φ_o

C'_o Total capacitance of C_H and C_o in series for ideal case

C_o Oxide capacitance

C_d Diffusion layer capacitance

C_T Total capacitance for MIE

C_H Helmholtz capacitance for MIE

E Electric field

E_{bd} Break-down electric field

K Boltzmann constant

n_o Concentration of electrolyte

Q Electron charge

T Temperature

V_{eo} Electro-osmotic velocity

V_G Voltage between metal and electrolyte

V_d Voltage between cathode and anode

V_{ss} Saturated surface potential

V_{pzz} Voltage V_G at which zeta potential is zero

V_{zs} Zero surface-charge potential

X_{ox} Oxide thickness

Z_{ie} Total impedance of double layer in parallel with Ca and the Warburg impedance Z_w

Greek alphabet

ε_e Dielectric constant of electrolyte

ε_o Absolute dielectric constant of vacuum

ε_r Relative dielectric constant

ξ Zeta potential

η Viscosity of the liquid

ρ Charge density

σ_m Charge at the metal-insulator interface

σ_o Surface charge at the insulator electrolyte due to specific adsorption

σ_B Charge at inner Helmhotz layer

σ_d Charge in the diffuse layer

φ_o Potential at the insulator-electrolyte interface

φ_B Potential at the inner Helmholtz layer

φ_d Potential at the outer Helholtz layer

6. Conclusion

A novel effect has been postulated by the name of field-effect electro-osmosis. The effect can change the electro-osmosis flow from left to right or from right to left, or it can make electro-osmosis zero.

Electro-osmosis is an electrokinetic phenomenon.

Author details

Kiumars Ghowsi[1*], Hosein Ghowsi[2] and M. Razazie [3]

*Address all correspondence to: k_ghowsi@yahoo.com

1 Department of Chemistry, Majlesi Branch, Islamic Azad University, Isfahan, Iran

2 Department of Mathematics, Payame Noor University, Tehran, Iran

3 Department of Material, Majlesi Branch, Islamic Azad University, Isfahan, Iran

References

[1] Nicholson and Carlisle, I. Nat. Philos., 4, 179 (1800); cited by H.A. Abramson, Electrokinetic Phenomena and Their Application to Biology and Medicine, Chemical Catalog Company, New York 1934.

[2] F.F. Reuss, Memoires de la Society Imperiale des Naturalites de Moscou, 2, 327 (1809).

[3] H. Helmholtz, Wied. Ann., 97, 337 (1879).

[4] K. Ghowsi, R.J. Gale, J. Chromatogr., 559, 95 (1991).

[5] V. Saxen, Wied. Ann., 47, 46 (1892).

[6] H. Freundlich, Kapillrchemic, Akademische Verlags Gesellschaft, Leipzig (1909).

[7] W.J. Siu, R.S.C. Cobbold, IEEE Trans. Electron Devices. Vol. Ed. -26, 1805-1851 (1979).

[8] J.O.M. Bockris and A.K.N. Keddy, Modern Electrochemistry, Vols. 1 and 2, New York, Plenum Press (1970).

[9] J. Jorgenson and K.D. Lukacs, Anal. Chem., 53, 1298 (1981).

[10] S. Terabe and T. Ando, Anal. Chem., 57, 834 (1985).

[11] K. Ghowsi Electrophoresis, Intech 1012 Chapter [online] Available at: www.intechopen.com

[12] K. Ghowsi, R.J. Gale, Biosensor Technology, Buck et al. (eds.) (1990).

Section 2

Capillary Electrophoresis

Chapter 2

Use of Amino Acid-Based Ionic Liquids in Capillary Electrophoresis

Constantina P. Kapnissi-Christodoulou

Additional information is available at the end of the chapter

http://dx.doi.org/10.5772/59663

1. Introduction

Ionic liquids (ILs) are unique solvents with melting points at or below 100 ºC. They have drawn scientific interest due to their unique properties that involve good thermal stability, miscibility in different solvents, tunable viscosity, conductivity, negligible vapor pressure, non-flammability, and low toxicity. They have often been called *designer solvents* in order to indicate their large structural variability due to the cation or the anion of the salt. This, in turn, broadens their application area (chemical reactions, catalysis, electrochemistry, separation, etc.) [1-5].

In the last few years, a big number of chiral ILs (CILs), have been designed, synthesized and used for applications in electrophoretic and chromatographic chiral discrimination [4, 6]. They play a key role in enantioselective analysis because they combine the advantages of ILs with the properties of a chiral moiety, which can be anionic and/or cationic. Their utility in separation science as chiral selectors, additives, chiral ligands, and chiral stationary phases is becoming increasingly important [7].

Although a large number of reviews have been provided on the synthesis of CILs [8-10], a surprisingly limited number of articles have been published on their utility in analytical separations, and particularly in electrophoretic separations [7, 11]. Capillary electrophoresis (CE) has been extensively used in chiral separations by using various chiral selectors, such as cyclodextrins (CDs), cyclofructans, oligo-and polysaccharides, polymeric surfactants, and others [12]. Some of the problems that limit their use involve low solubility, instability at high temperatures and/or low pH values, time-consuming organic synthesis procedures and high cost. The use of CILs is considered a potential alternative because they can dissolve various polar and nonpolar analytes, they may provide chiral selectivity, and their synthesis procedure is simple. In CE, the CILs are mainly used as BGE additives, and secondarily as chiral ligands and chiral selectors.

A new class of CILs, called amino acid ester-based ILs (AAILs), was synthesized and characterized in 2005 by Tao *et al.* [13]. AAILs consist of cations, which are derived from amino acids or amino acid esters, such as glycine (Gly), alanine (Ala), alanine methyl ester ($AlaC_1$), and alanine ethyl ester ($AlaC_2$), and commonly used anions, such as nitrate (NO_3), tetrafluoroborate (BF_4), lactate (Lac) and bis(trifluromethane)sulfonamide (NTf_2). Bwambok *et al.* [14] used a simple metathesis reaction to synthesize L-and D-alanine *tert* butyl ester-based ILs with several anions, and they studied their enantiomeric recognition properties by using fluorescence and nuclear magnetic resonance spectroscopy.

In this chapter, the ability of AAILs to be used as chiral and achiral media in CE is investigated. In particular, some representative studies that involve the utility of AAILs as background electrolyte (BGE) additives and as sole chiral selectors in electrophoretic separations are reported and discussed. These studies involve synthesis procedure, establishment of optimum separation conditions and method validation. The first part of this chapter involves the application of AAILs as BGE additives and the evaluation of their performance in both chiral and achiral analysis, while the second part demonstrates their chiral recognition ability for the enantioseparation of 1,1′-binaphthyl-2,2-diylhydrogenphosphate (BNP).

2. AAILs as background electrolyte additives

Most of the applications of AAILs (Figure 1) in electrophoretic separations have been as additives in BGEs [15-23]. Hadjistasi *et al.* [15] were the first to report the use of an AAIL as an additive in the BGE to improve electrophoretic separations. In particular, the addition of the CIL D-alanine *tert*-butyl ester lactate (D-$AlaC_4$Lac) into the BGE improved the resolution of the pipecolic acid enantiomers. This CIL provided an increase in resolution from 1.41, obtained when β-cyclodextrin (β-CD) was used as the sole chiral selector, to 1.87, obtained when both chiral selectors were added in the BGE. In 2013, two novel AAILs, tetramethylammonium-L-arginine (TMA-L-Arg) and tetramethylammonium-L-aspartic acid (TMA-L-Asp), were applied, for the first time, in CE, with glycogen as the chiral selector, in order to evaluate their potential synergistic effect [16]. Glycogen is an electrically neutral and branched polysaccharide that has, over the years, proven its chiral recognition abilities for the enantioseparation of various basic and acidic drug compounds. In their study, it was observed that, when the TMA-L-Arg/glycogen and TMA-L-Asp/glycogen systems were applied, both resolution and selectivity were significantly improved in comparison to the single glycogen separation system. This, in turn, suggested the existence of the synergistic effect. In a more recent study, the vancomycin-based synergistic system with L-alanine and L-valine *tert* butyl ester bis(trifluoromethane)sulfonamide (L-$AlaC_4NTf_2$ and L-$ValC_4NTf_2$) as additives was evaluated in CE for the enantioseparation of five profens [18]. When the binary systems were applied, all enantioseparations were considerably improved, and the resolution values were greatly higher than in the case where vancomycin was used as the sole chiral selector. In this part of the chapter, two representative studies performed recently in chiral and achiral CE are briefly described [17, 22].

http://dx.doi.org/10.5772/59663

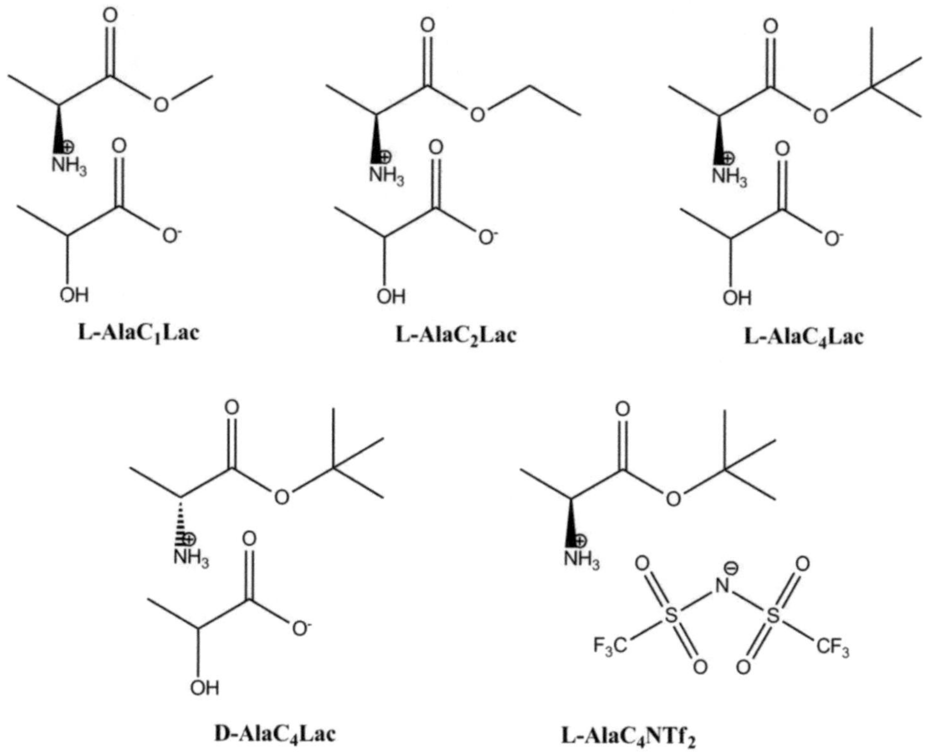

Figure 1. Structures of the AAILs L-AlaC_1Lac, L-AlaC_2Lac, L-and D-AlaC_4Lac, and L-AlaC_4NTf_2 [27].

2.1. Chiral analysis

The use of AAILs as BGE additives, for improved resolutions, selectivity factors, and efficiencies in chiral analysis, is demonstrated further here by providing a more in-depth analysis of a research work that was performed by Zhang *et al.* [17]. In their study, two AAILs (L-AlaC_4NTf_2 and L-ValC_4NTf_2) were applied as additives and β-CD derivatives (methyl-β-CD, hydropropyl-β-CD, glucose-β-CD) as chiral selectors in CE for the enantioseparation of six anionic racemic drug compounds.

The synthesis of both AAILs was accomplished by use of a one-step anion exchange reaction of the corresponding amino acid ester chloride and the bis(trifluoromethane)sulfonimide lithium salt [14]. Briefly, an appropriate amount of L-alanine and L-valine *tert* butyl ester hydrochloride and an equimolar amount of bis(trifluoromethane)sulfonimide lithium salt were separately dissolved in distilled water. Then, the two solutions were mixed and stirred for 2 h at room temperature. The mixture resulted in two layers, of which the lower layer was separated and dried under vacuum overnight. The resulted products were colorless liquids.

The main objective of their study was to evaluate the synergistic effect of the AAILs with the β-CD derivatives. It was proven to be significant for half of the analytes examined, and particularly for naproxen, pranoprofen and warfarin. Figure 2 demonstrates the electropherograms obtained when β-CD derivatives were used as the sole chiral selectors [(a)], and when AAILs were used as additives [(b) and (c)].

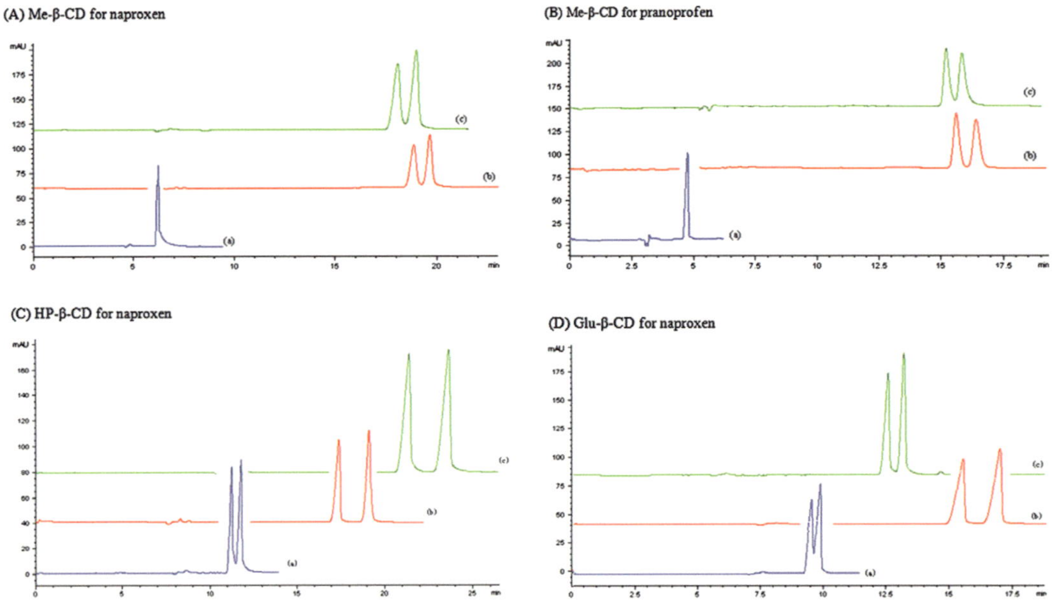

Figure 2. Electropherograms with the absence and presence of AAILs for enantioseparation with (A) methyl-β-CD for naproxen, (B) methyl-β-CD for pranoprofen, (C) hydropropyl-β-CD for naproxen and (D) glucose-β-CD for naproxen. Conditions: fused-silica capillary, 33 cm (24.5 cm effective length) × 50 μm id; 30 mM sodium citrate/citric acid buffer solution with 20% (v/v) ethanol for naproxen or 20% (v/v) acetonitrile for pranoprofen, containing (A) and (B) (a) methyl-β-CD; (b) 20 methyl-β-CD+15 mM L-AlaC$_4$NTf$_2$; (c) 20 mM methyl-β-CD+15 mM L-ValC$_4$NTf$_2$; (C) (a) 30 mM hydropropyl-β-CD; (b) 30 mM hydropropyl-β-CD+15 mM L-AlaC$_4$NTf$_2$; (c) 30 mM hydropropyl-β-CD+15 mM L-ValC$_4$NTf$_2$, (D) (a) 30 mM glucose-β-CD; (b) 30 mM glucose-β-CD+15 mM L-AlaC$_4$NTf$_2$; (c) 30 mM glucose-β-CD+15 mM L-ValC$_4$NTf$_2$; pH 5.0; applied voltage, 20 kV; capillary temperature, 25°C [17].

The novel synergistic system was optimized by using methyl-β-CD/AAILs as model systems. An important factor affecting the enantioseparation is the concentration of the chiral selector and the CIL, since both concentrations will determine the equilibria between the chiral selector, the CIL and the enantiomers. Initially, at a fixed concentration of 20 mM methyl-β-CD, as the concentrations of L-AlaC$_4$NTf$_2$ and L-ValC$_4$NTf$_2$ increased from 5 to 15 mM both resolution and effective selectivity factor improved. However, at the concentration of 20 mM, no peak was observed, even at 60 min. The electrophoretic mobility decreased dramatically, possibly due to the adsorption of the CIL cations onto the capillary walls.

The effect of the chiral selector's concentration on enantioseparation was also studied by varying the concentration of methyl-β-CD from 10 to 50 mM. In the single methyl-β-CD separation system, resolution values increased upon increasing the concentration, due to an increase in the complexation between the chiral selector and the enantiomers. On the other hand, in the methyl-β-CD/AAILs systems, resolution was initially increased, and then decreased mainly due to the gradual complex saturation. Therefore, the optimum concentration of methyl-β-CD is lower with the presence of the synergistic effect in comparison to the single methyl-β-CD system.

The optimum BGE composition was also determined by examining two different systems (sodium sodium acetate-acetic acid and citrate-citric acid). The first BGE provided deformed peaks, while the use of the second one resulted in better peak shapes and better resolutions. The BGE pH was also investigated for the chiral recognition process, because the degree of protonation in the analytes and the AAILs depends on this parameter. It was observed that at pH values below 4.4 or above 5.6, resolutions of all the understudy chiral compounds were reduced. Therefore, a pH value of 5.0 was considered the optimum. According to the authors, at the particular optimum pH, in the present system of analyte/AAIL/methyl-β-CD, the following bindings occur: (a) hydrogen bonding among the hydroxyl function in methyl-β-CD, carboxyl group in the drug compounds and amino function in the AAILs and (b) ionic interactions between the carboxyl group in the analytes and the amino function in the AAILs.

Another important observation, in this study, was the improvement of both resolution and effective selectivity factor with the addition of an organic modifier, possibly due to a decrease in electroosmotic mobility, which, in turn, increases the interactions between the AAIL, methyl-β-CD and analyte. Other parameters, such as composition and pH of buffer system and applied voltage were also examined, and the optimum conditions included 15 mM AAIL, 20 mM methyl-β-CD, 30 mM sodium citrate/citric acid (pH 5), and 20 kV.

2.2. Achiral analysis

In a different study, the synergistic effect of sodium dodecyl sulfate (SDS) and L-alanine *tert* butyl ester lactate (L-AlaC$_4$Lac) was examined by Mavroudi *et al.* [22] for the separation of 2-arylpropionic acid non-steroidal anti-inflammatory drugs (Figure 3). The BGE was supported with either SDS, or an AAIL or SDS/AAIL, and their performance was evaluated by comparing migration times, efficiencies and %RSD values. Many analytical CE methods have, so far, been used for the separation of a wide variety of NSAIDs, by applying different modes of CE. These include micellar electrokinetic chromatography (MEKC) with SDS as an additive, capillary electrochromatography (CEC) with poly(stearyl methacrylate-divinyl benzene) monolithic columns and non-aqueous CE [24-26]. In the study reported by Maria *et al.*, the AAIL L-AlaC$_4$Lac was applied in CE as a sole additive for the simultaneous separation of NSAIDs for the first time. A very important consideration in this study was whether the use of an AAIL as a sole additive resulted in more effective separations than in the case of using a common surfactant. Therefore, a comparison was performed by adding L-AlaC$_4$Lac or SDS into the BGE (100 mM Tris/10 mM tetraborate decahydrate).

For the synthesis of L-AlaC$_4$Lac, appropriate amounts of the corresponding amino acid ester hydrochloride and silver lactate were separately mixed with methanol. The amino acid ester solution and the suspended silver lactate were then mixed and stirred. Subsequently, the precipitate was filtered and removed, and the remained solution was evaporated *in vacuo* and purified by being crystallized in methanol/ether [14].

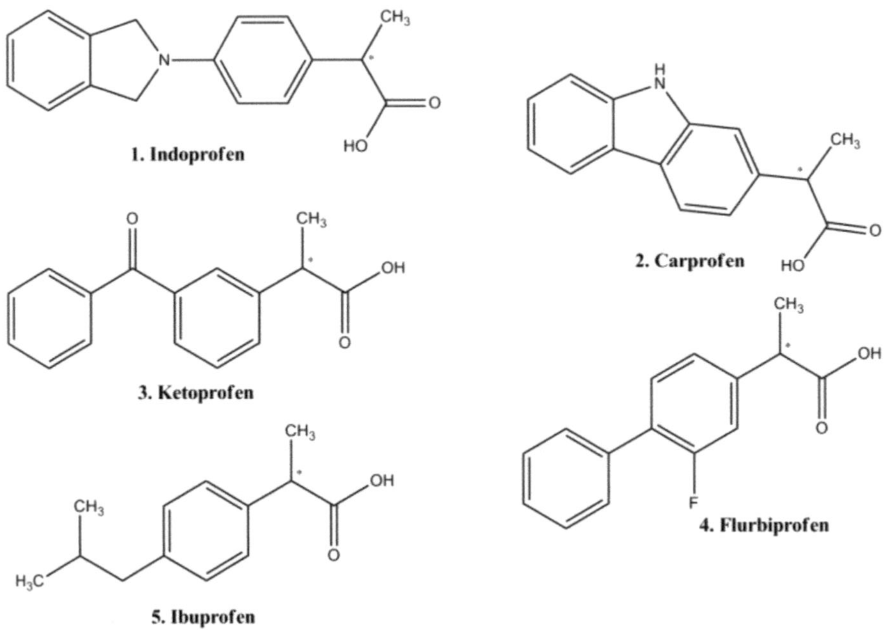

Figure 3. Structures of the five NSAIDs used in this study.

In the single SDS system, separation is achieved by differential partitioning of analytes between the hydrophilic core of the surfactant and the bulk aqueous phase via electrostatic and hydrophobic interactions, and hydrogen bonding. SDS was initially examined as a sole additive, by varying its concentration from 10 to 30 mM. It was observed from Figure 4 that, even though the migration times of all analytes increased along with the increase of SDS concentration, resolution values decreased dramatically. At 30 mM, a coelution of indoprofen and ketoprofen was observed, and carprofen did not elute, even at 30 min. This is because the predominant population, at pH 8, is in the anionic form, which is expected to be repelled by the negatively charged headgroup of the anionic surfactant.

For comparison purposes and for further optimization of the separation, an AAIL was used as an additive. The effect of its concentration on the separation of NSAIDs was first examined. The optimum concentration was determined according to resolution, efficiency and analysis time. As demonstrated in Figure 5, when a 20-mM concentration was used, the Rs values for the peak pairs carprofen-ketoprofen and flurbiprofen-ibuprofen were 1.3 and 1.2, respectively. Concentrations above 30 mM provided baseline separations with Rs values higher than 1.5. In addition, the total analysis time, in the cases of 20, 30 and 40 mM L-AlaC$_4$Lac, was not altered (~ 9.5 min), while, from 50 to 70 mM, it was increased to ~ 12 min. Efficiency was determined by calculating the number of theoretical plates (N) for all peaks. It was observed that at a concentration of 40 mM, N for all peaks was very high, in comparison to the ones obtained when the other concentrations were applied (Figure 6). Another important observation involved the elution order of NSAIDs, which was different from the elution order observed when the SDS was added in the BGE, probably due to the different types of interactions

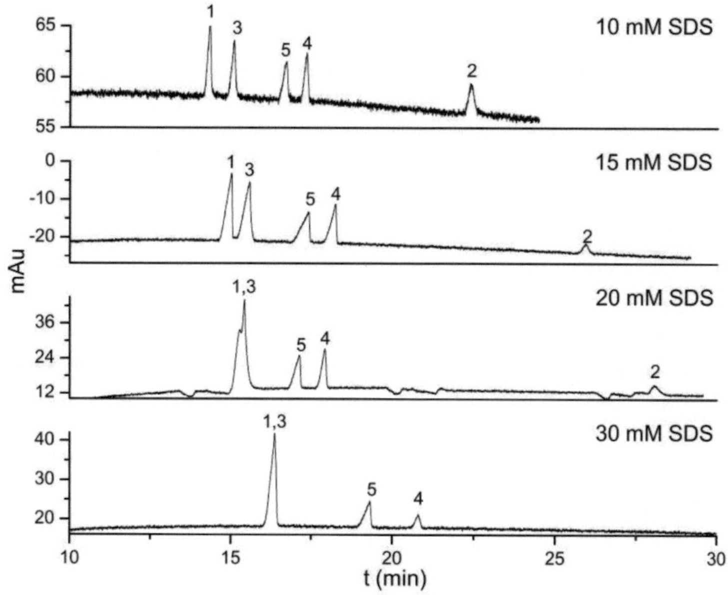

Figure 4. Effect of SDS concentration on the simultaneous separation of NSAIDs. Conditions: 100 mM Tris/10 mM tetraborate decahydrate pH 8, applied voltage 30 kV, temperature 20 °C, detection wavelength 200 nm [22].

between the additive and the analytes. In the case of SDS, the interactions are based on hydrophobicity, while in the case of the additive L-AlaC_4Lac, the separation is based on electrostatic interactions.

pH is another important parameter that is necessary to be optimized because, firstly, alterations in pH can affect the analyte charge and, secondly, the primary amine group of the cation of the AAIL can be positively charged or neutral in several pHs. An increase in pH from 8.0 to 8.5, the resolution decreased from 2.9 to 1.5 for peaks carprofen-ketoprofen and from 1.7 to 1.3 for peaks flurbiprofen-ibuprofen. A further increase in pH (9.0 and 9.5) resulted in two coelutions and shorter migration times due to fewer electrostatic interactions between the AAIL and the negatively charged analyte, since the amount of the positively charged amino group is decreased. In addition, at high pH values, the AAIL may undergo ester hydrolysis, which results in the lack of the *tert* butyl group in the cation.

The reproducibilities were also evaluated and compared by calculating the relative standard deviation (RSD) values of the electroosmotic flow (EOF) and the migration times of all the analyte peaks. In particular, in both SDS and L-AlaC_4Lac cases, the run-to-run RSD values were obtained from 10 consecutive electrophoresis runs. In the case of SDS, the RSD of the EOF was 2.1%, and the RSD values of the analytes ranged from 2.8% to 11.7%. In the case of L-AlaC_4Lac, the RSD of the EOF was 0.4% and the RSDs of the NSAIDs ranged from 1.2% to 1.3% (Table 1). In the same table, a comparison between the two additives in regard to efficiency is also demonstrated. The efficiency of all analyte peaks was above 102,000 for L-AlaC_4Lac, in comparison to SDS, which provided efficiency values between 47,000 and 76,000 theoretical plates.

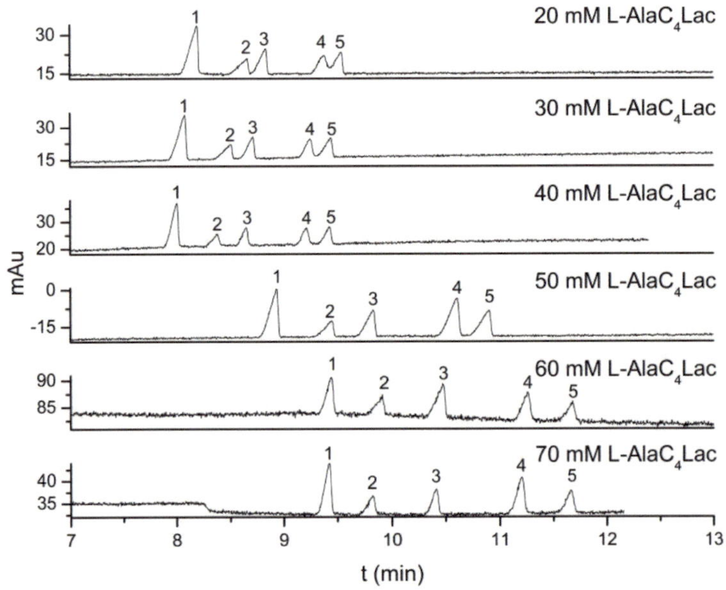

Figure 5. Effect of L-AlaC$_4$Lac concentration on the simultaneous separation of NSAIDs. Conditions: 100 mM Tris/10 mM tetraborate decahydrate pH 8, applied voltage 30 kV, temperature 35 °C, detection wavelength 200 nm [22].

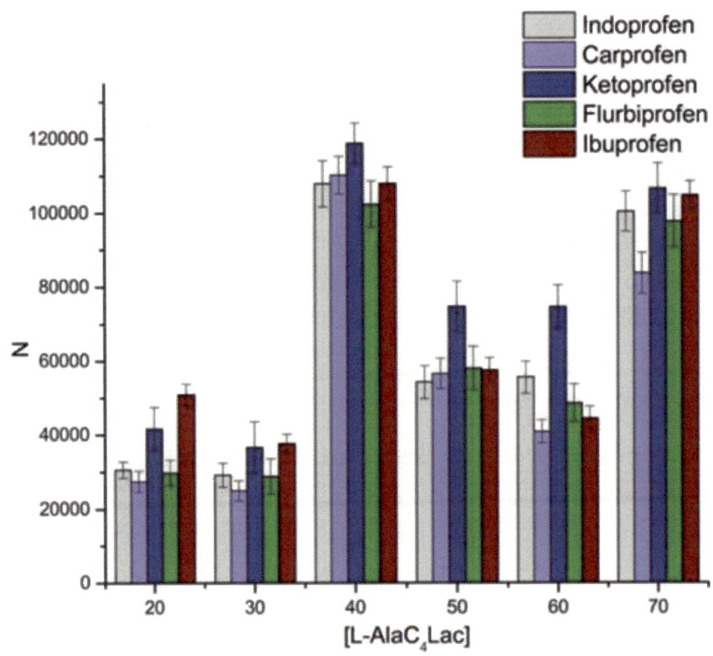

Figure 6. Effect of L-AlaC$_4$Lac concentration on N. Conditions: 100 mM Tris/10 mM tetraborate decahydrate pH 8, applied voltage 30 kV, temperature 35 °C, detection wavelength 200 nm [22].

http://dx.doi.org/10.5772/59663

		10 mM SDS			40 mM L-AlaC$_4$Lac		
Peak	**Analyte**	**t (min)**	**RSD %**	**N**	**t (min)**	**RSD %**	**N**
EOF	—	5.50	2.1	—	3.69	0.4	—
1	**Indoprofen**	14.33	3.0	64,722	8.03	1.2	107,863
2	**Carprofen**	22.39	11.7	51,868	8.48	1.2	110,156
3	**Ketoprofen**	15.09	2.8	60,805	8.78	1.3	118,785
4	**Flurbiprofen**	17.34	4.1	75,824	9.43	1.2	102,357
5	**Ibuprofen**	16.71	3.6	46,968	9.65	1.3	107,943

Table 1. Run-to-run reproducibility and efficiency of SDS and L-AlaC$_4$Lac for the simultaneous separation of NSAIDs. Conditions: 100 mM Tris/10 mM tetraborate decahydrate pH 8, applied voltage 30 kV, temperature 20 °C (for SDS) and 35 °C (for L-AlaC$_4$Lac), detection wavelength 200 nm [22].

A last consideration for this study involved the effect of the addition of both SDS and L-AlaC$_4$Lac into the BGE on the separation of NSAIDs. A concentration of 10 mM SDS and different concentrations of L-AlaC$_4$Lac were added into the BGE (1-40 mM). An increase in the concentration of L-AlaC$_4$Lac resulted in a more effective separation of NSAIDs, in regard to efficiency and resolution, probably due to the synergistic effect of the SDS/L-AlaC$_4$Lac system. However, the elution order and analysis time (~ 23 min) were similar to the ones obtained when the SDS was used as the sole additive. Therefore, it was clear from this study that the additive L-AlaC$_4$Lac is considered an effective alternative to SDS for a reproducible, baseline, high efficient and fast separation of NSAIDs.

3. AAILs as chiral selectors

Although numerous studies reported the use of AAILs in electrophoretic enantiomeric separation, only one study was performed by using AAILs both as co-electrolytes and chiral selectors [27]. However, other CILs, that are not amino-acid based, have been added into the BGE and used as sole chiral selectors for the enantioseparation of a number of analytes [28-31]. Yuan *et al.*[28] used the IL (R)-N,N,N-trimethyl-2-aminobutanol-bis(trifluoromethanesulfon)imidate as the sole chiral selector in CE, GC and HPLC for the separation of fifteen enantiomeric compounds. It was used as an additive in a BGE of 20 mM Na_2HPO_4-NaH_2PO_4 or 20 mM $Na_2B_4O_7$. The resolution values varied from 0.60 for di-O,O′ p-toluyl-tartaric acid to 6.80 for 3-benzyloxy-1,2-propane diol.

In another study by Tran and Mejac [29], the CIL *S*-[3-(chloro-2-hydroxypropyl)trimethylammonium] [bis((trifluoromethyl)sulfonyl)amide] (*S*-[CHTA]$^+$[Tf$_2$N]$^-$) was utilized as a sole chiral selector for the enantioseparation of a number of pharmaceutical compounds. Even though *S*-$[CHTA]^+[Tf_2N]^-$can serve as a chiral selector, it was unable to be used as a sole chiral selector since no enantioseparation could be achieved. Therefore, more BGE chiral additives were used, such as the chiral anion sodium cholate and the neutral chiral 1-*S*-octyl-β-D-thioglucopyranoside (OTG).

Ma *et al.* [30] explored the potential of using an ephedrine-based CIL, (+)-*N*,*N*-dimethylephedrinium-bis(trifluoromethanesulfon)imidate ($[DMP]^+[Tf_2N]^-$) as both a chiral selector and a BGE in nonaqueous CE. The addition of $[DMP]^+[Tf_2N]^-$resulted in a reversed EOF (anodic flow); so, the experiments were performed in the reversed polarity mode by using acetone as the EOF marker. The enantioseparations of rabeprazole and omeprazole were achieved mainly due to the different ion-pair formation equilibrium constants between the ephedrine-based CIL cations and the negatively charged enantiomers, and the hydrogen bonding, and secondarily due to other interactions, such as π-π interactions and dipole-dipole interactions.

In 2013, Yu *et al.* [31] synthesized a novel CIL functionalized β-CD (6-*O*-2-hydroxypropyltrimethylammonium-β-CD tetrafluoroborate, [HPTMA-β-CD][BF_4]) and applied it as a chiral selector in CE for the enantioseparation of eight chiral drug compounds. The separation conditions were optimized by studying the effect of the CIL concentration and the BGE pH. The results obtained demonstrated the excellent chiral discriminating ability of [HPTMA-β-CD][BF_4].

As mentioned earlier, the use of AAILs as sole chiral selectors has only been reported once by Stavrou *et al.* [27]. Their applications are demonstrated further here by providing a more in-depth and informative CE chiral analysis of BNP. In their study, five AAILs [L-alanine methyl ester lactate, L-alanine ethyl ester lactate, L-and D-alanine *tert* butyl ester lactate (L-AlaC$_1$Lac, L-AlaC$_2$Lac and L-and D-AlaC$_4$Lac), and L-AlaC$_4$NTf$_2$] (Figure 1) were synthesized and used as additives in the BGE in order to evaluate their chiral recognition ability by comparing the resolution values.

The optimum separation conditions were established by altering different important parameters, such as the alkyl ester group, the anion, the configuration and the concentration of the AAIL. In their first study, the influence of steric hindrance on the enantiomeric separation of BNP was examined by applying separately as sole chiral selectors the AAILs L-AlaC$_1$Lac, L-AlaC$_2$Lac and L-AlaC$_4$Lac at concentrations of 60 mM and 100 mM. It was observed that as the length and the bulkiness of the ester group increased, the resolution of BNP increased. In particular, the first AAIL did not demonstrate any enantioselectivity, while the second one was able to provide partial enantioseparation (R_S: 1.09). However, when L-AlaC$_4$Lac was used at both concentrations, a baseline separation was achieved with R_S values of 1.94 and 2.43 (Table 2). It is, therefore, concluded that the enantioseparation of BNP is favored in the presence of *tert*-butyl group, and steric hindrance is involved in the enantioseparation mechanism.

Two very important considerations in this study involved the effect of the anion and the configuration of the cation on resolution. Two different anions were used (Lac and NTf$_2$), which provided baseline separation (Table 2). In particular, the resolution obtained by use of L-AlaC$_4$NTf$_2$ was slightly lower (R_S=1.72) than the one obtained with L-AlaC$_4$Lac (R_S=1.94),

http://dx.doi.org/10.5772/59663

possibly due to the low solubility of the first in water, which provides fewer free cations and less interaction with the analyte molecules. In addition, the results obtained with NTf_2 were not reproducible and provided an unstable baseline. As far as the cation configuration is concerned, the D-and L-$AlaC_4Lac$ were used at a concentration of 60 mM in order to compare their enantioseparation ability. It was observed that the R_S values obtained were very similar (1.95 and 1.94), while the elution order was different. As expected, the chiral center of the cation is considered the main active center of the chiral selector. Therefore, the elution order of the enantiomers of the analyte is based on the configuration of the CIL.

CIL	*[CIL]	t_{EOF}	t_1	t_2	R_S
L-$AlaC_2Lac$	60 mM	5.755	11.471	11.511	-
	100 mM	6.272	13.929	14.065	1.09
L-$AlaC_4Lac$	60 mM	6.207	12.702	12.922	1.94
	100 mM	6.650	13.064	13.377	2.43
D-$AlaC_4Lac$	60 mM	6.172	12.499	12.702	1.95
L-$AlaC_4NTf_2$	60 mM	6.495	14.622	14.865	1.72

Table 2. Effect of the cation, the anion and the configuration of the CIL on R_S. Conditions: BGE, 100 mM Tris/10 mM Borate pH=8; applied voltage, 30 kV; capillary temperature, 25 °C; detection wavelength, 214 nm.

In another study for the optimization of the BNP enantioseparation, the effect of the concentration of L-$AlaC_4Lac$ on resolution was investigated. As demonstrated in Figure 7, resolution increased significantly from ~ 0.4 to 2.43 with increasing the concentration from 20 to 100 mM. It was also observed that an increase in the AAIL concentration resulted in a decrease in the electroosmotic flow mobility, probably due to the coating of the AAIL cations onto the capillary wall Figure 8. The electrophoretic mobility, μ_{EOF}, was calculated according to:

$$\mu_{EOF} = \frac{lL}{t_{nm}V}$$

where L is the total capillary length, l is the effective capillary length (from the injection end to the detector), t_{nm} is the migration time of the neutral marker and V is the applied voltage.

The last parameter examined was the BGE pH, which was important, since the cation of the CILs used seems to be pH depended. Resolution decreased from 1.94 to 1.29, upon increasing the pH from 8 to 8.5, while at higher pHs (9 and 10), no enantioseparation was observed. As mentioned in Section 2.2, an increase in the pH decreases the amount of the positively charged amino group, which consequently reduces the electrostatic interactions between the AAIL and the negatively charged analyte. In addition, at high pH values, and particularly pH 9 and 10, the AAIL may undergo ester hydrolysis. This results in the lack of the *tert* butyl group in the cation, which is an important factor for the particular enantioseparation. Finally, according to the authors, it is concluded from the above-mentioned studies that the enantioseparation

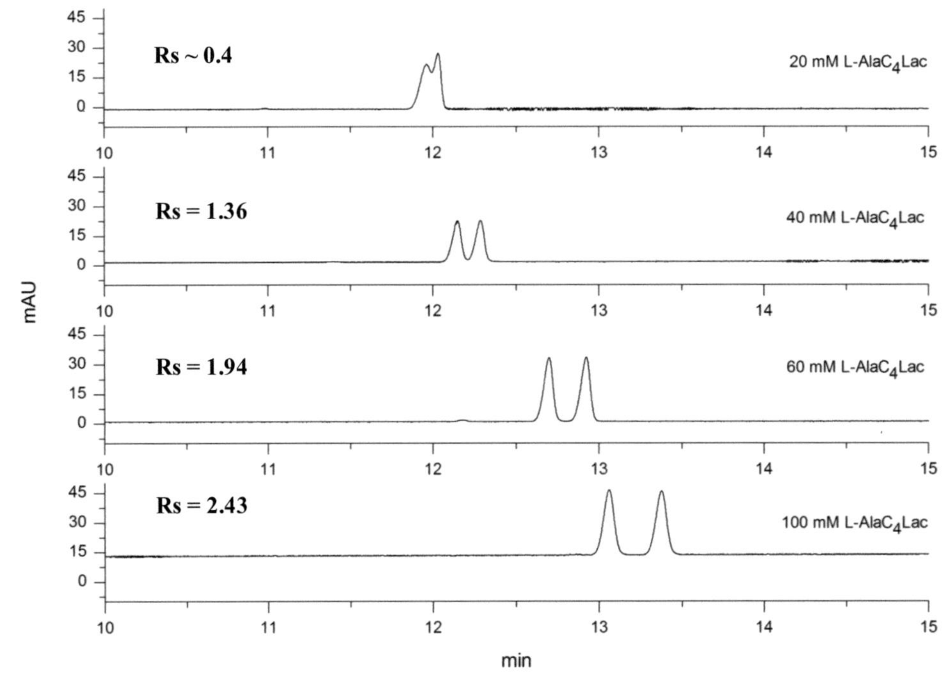

Figure 7. Effect of L-AlaC$_4$Lac concentration on the enantioseparation of BNP. Conditions: BGE, 100 mM Tris/10 mM Borate pH=8 applied voltage, 30 kV; capillary temperature, 25 °C; detection wavelength, 214 nm [27].

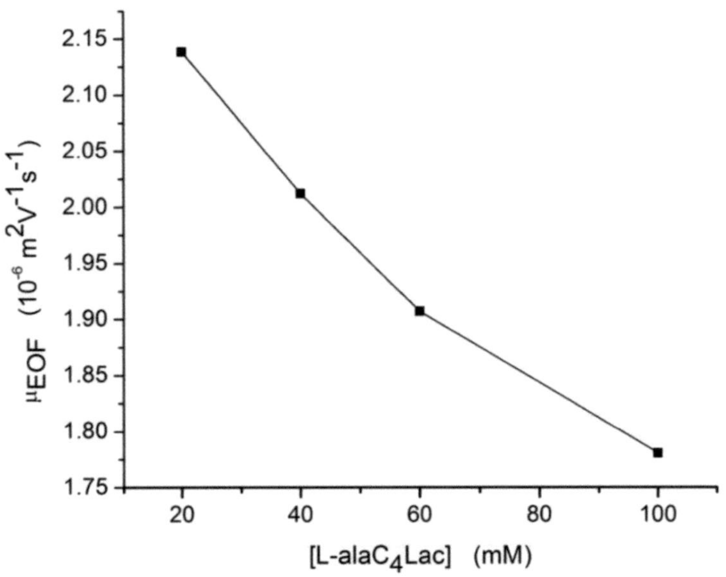

Figure 8. Effect L-AlaC$_4$Lac concentration on μ_{EOF}. Conditions: BGE, 100 mM Tris/10 mM Borate pH=8 applied voltage, 30 kV; capillary temperature, 25 °C; detection wavelength, 214 nm [27].

mechanism for this particular application is based on: (a) steric hindrance (*tert* butyl group), (b) electrostatic interactions (between the cation of the CIL and the negatively charged analyte) and (c) hydrogen bonding (hydrogen-bonding capability of the phosphate group in BNP).

4. Concluding remarks

In this chapter, the suitability of the AAILs in chiral and achiral CE analysis was evaluated. These new AAILs, which can be easily synthesized from commercially available reagents, proved to be efficient chiral additives for the enantioseparation of different analytes. Even though only a limited number of studies have, so far, applied AAILs as BGE additives and chiral selectors, it is easy to conclude from the data demonstrated in this chapter that the future of AAILs in separation science has a great deal of potential, and it is expected to expand significantly. Further research though is required in order to understand the chiral recognition mechanisms between the AAILs, the common chiral selectors and the enantiomers. This will, in turn, help us design even more effective AAILs for applications in chiral electrophoretic and chromatographic recognition.

Author details

Constantina P. Kapnissi-Christodoulou*

Address all correspondence to: ckapni1@ucy.ac.cy

Department of Chemistry, University of Cyprus, Nicosia, Cyprus

References

[1] Wasserscheid, P., Keim, W. (2000) Ionic liquids-New 'solutions' for transition metal catalysis. Angewandte Chemie-International Edition 39: 3773-3789.

[2] Dupont, J., De Souza, R. F., Suarez, P. A. Z. (2002) Ionic liquid (molten salt) phase organometallic catalysis. Chemical Reviews 102: 3667-3692.

[3] Jain, N., Kumar, A., Chauhan, S., Chauhan, S. M. S. (2005) Chemical and biochemical transformations in ionic liquids. Tetrahedron 61: 1015-1060.

[4] Chen, X., Li, X., Hu, A., Wang, F. (2008) Advances in chiral ionic liquids derived from natural amino acids. Tetrahedron: *Asymmetry* 19: 1-14.

[5] Liu, J. F., Jiang, G. B., Jönsson, J. A. (2005) Application of ionic liquids in analytical chemistry. TrAC-Trends in Analytical Chemistry 24: 20-27.

[6] Kapnissi-Christodoulou, C. P., Stavrou, I. J., Mavroudi, M. C. (2014) Chiral ionic liquids in chromatographic and electrophoretic separations. Journal of Chromatography A In Press.

[7] Payagala, T., Armstrong, D. W. (2012) Chiral ionic liquids: A compendium of syntheses and applications (2005-2012). Chirality 24: 17-53.

[8] Pereira, M. M. A. (2012) Chiral ionic liquids from Carbohydrates: Synthesis and properties. Mini-Reviews in Organic Chemistry 9: 243-260.

[9] Tan, Z. Q., Liu, J. F., Pang, L. (2012) Advances in analytical chemistry using the unique properties of ionic liquids. TrAC-Trends in Analytical Chemistry 39: 218-227.

[10] Winkel, A., Reddy, P. V. G., Wilhelm, R. (2008) Recent advances in the synthesis and application of chiral ionic liquids. Synthesis 999-1016.

[11] Shamsi, S. A., Danielson, N. D. (2007) Utility of ionic liquids in analytical separations. Journal of Separation Science 30: 1729-1750.

[12] Tsioupi, D. A., Stefan-van Staden, R. I., Kapnissi-Christodoulou, C. P. (2013) Chiral selectors in CE: Recent developments and applications. Electrophoresis 34: 178-204.

[13] Tao, G. H., He, L., Sun, N., Kou, Y. (2005) New generation ionic liquids: Cations derived from amino acids. Chemical Communications 3562-3564.

[14] Bwambok, D. K., Marwani, H. M., Fernand, V. E., Fakayode, S. O., Lowry, M., Negulescu, I., Strongin, R. M., Warner, I. M. (2008) Synthesis and characterization of novel chiral ionic liquids and investigation of their enantiomeric recognition properties. Chirality 20: 151-158.

[15] Hadjistasi, C. A., Stavrou, I. J., Stefan-Van Staden, R. I., Aboul-Enein, H. Y., Kapnissi-Christodoulou, C. P. (2013) Chiral separation of the clinically important compounds fucose and pipecolic acid using ce: Determination of the most effective chiral selector. Chirality 25: 556-560.

[16] Zhang, Q., Du, Y. (2013) Evaluation of the enantioselectivity of glycogen-based synergistic system with amino acid chiral ionic liquids as additives in capillary electrophoresis. Journal of Chromatography A 1306: 97-103.

[17] Zhang, J., Du, Y., Zhang, Q., Chen, J., Xu, G., Yu, T., Hua, X. (2013) Investigation of the synergistic effect with amino acid-derived chiral ionic liquids as additives for enantiomeric separation in capillary electrophoresis. Journal of Chromatography A 1316: 119-126.

[18] Zhang, J., Du, Y., Zhang, Q., Lei, Y. (2014) Evaluation of vancomycin-based synergistic system with amino acid ester chiral ionic liquids as additives for enantioseparation of non-steroidal anti-inflamatory drugs by capillary electrophoresis. Talanta 119: 193-201.

[19] Zuo, L., Meng, H., Wu, J., Jiang, Z., Xu, S., Guo, X. (2013) Combined use of ionic liquid and β-CD for enantioseparation of 12 pharmaceuticals using CE. Journal of Separation Science 36: 517-523.

[20] Cui, Y., Ma, X., Zhao, M., Jiang, Z., Xu, S., Guo, X. (2013) Combined use of ionic liquid and hydroxypropyl-β-cyclodextrin for the enantioseparation of ten drugs by capillary electrophoresis. Chirality 25: 409-414.

[21] Zhao, M., Cui, Y., Yu, J., Xu, S., Guo, X. (2014) Combined use of hydroxypropyl-β-cyclodextrin and ionic liquids for the simultaneous enantioseparation of four azole antifungals by CE and a study of the synergistic effect. Journal of Separation Science 37: 151-157.

[22] Mavroudi, M. C., Kapnissi-Christodoulou, C. P. (2014) Evaluation of amino acid ester-based ionic liquids as buffer additives in CE for the separation of 2-arylpropionic acids nonsteroidal anti-inflammatory drugs. Electrophoresis

[23] Yujiao, W., Guoyan, W., Wenyan, Z., Hongfen, Z., Huanwang, J., Anjia, C. (2014) Chiral separation of phenylalanine and tryptophan by capillary electrophoresis using a mixture of β-CD and chiral ionic liquid ([TBA] [l-ASP]) as selectors. Biomedical Chromatography 28: 610-614.

[24] Macià, A., Borrull, F., Calull, M., Aguilar, C. (2006) Different sample stacking strategies to analyse some nonsteroidal anti-inflammatory drugs by micellar electrokinetic capillary chromatography in mineral waters. Journal of Chromatography A 1117: 234-245.

[25] Hsu, C. H., Cheng, Y. J., Singco, B., Huang, H. Y. (2011) Analyses of non-steroidal anti-inflammatory drugs by on-line concentration capillary electrochromatography using poly(stearyl methacrylate-divinylbenzene) monolithic columns. Journal of Chromatography A 1218: 350-358.

[26] Bonvin, G., Schappler, J., Rudaz, S. (2014) Non-aqueous capillary electrophoresis for the analysis of acidic compounds using negative electrospray ionization mass spectrometry. Journal of Chromatography A 1323: 163-173.

[27] Stavrou, I. J., Kapnissi-Christodoulou, C. P. (2013) Use of chiral amino acid ester-based ionic liquids as chiral selectors in CE. Electrophoresis 34: 524-530.

[28] Yuan, L. M., Han, Y., Zhou, Y., Meng, X., Li, Z., Zi, M., Chang, Y. (2006) (R)-N,N,N-trimethyl-2-aminobutanol-bis(trifluoromethane-sulfon)imidate chiral ionic liquid used as chiral selector in HPCE, HPLC, and CGC. Analytical Letters 39: 1439-1449.

[29] Tran, C. D., Mejac, I. (2008) Chiral ionic liquids for enantioseparation of pharmaceutical products by capillary electrophoresis. Journal of Chromatography A 1204: 204-209.

[30] Ma, Z., Zhang, L., Lin, L., Ji, P., Guo, X. (2010) Enantioseparation of rabeprazole and omeprazole by nonaqueous capillary electrophoresis with an ephedrine-based ionic liquid as the chiral selector. Biomedical Chromatography 24: 1332-1337.

[31] Yu, J., Zuo, L., Liu, H., Zhang, L., Guo, X. (2013) Synthesis and application of a chiral ionic liquid functionalized β-cyclodextrin as a chiral selector in capillary electrophoresis. Biomedical Chromatography 27: 1027-1033.

Chapter 3

Importance of Treatment Process on the Analysis of Penicillins in Milk Samples by Capillary Electrophoresis

J.A. Rodríguez, E. Martínez, I.S. Ibarra, J.M. Miranda,
C.A. Galán-Vidal and M.E. Páez-Hernández

Additional information is available at the end of the chapter

http://dx.doi.org/10.5772/59371

1. Introduction

Some veterinary drugs (such as antibiotics) are required in rational use at some stage of animal production in order to guarantee the hygienic management of farms. Antibiotics can be used at low levels for growth promotion (2.5 to 125 mg_{antib} $kg^{-1}{}_{feed}$, depending on the species treated), at intermediate levels for disease prevention (<200 mg_{antib} $kg^{-1}{}_{feed}$), and at high levels to treat infected animals. Prophylactic antimicrobial use is applied in intense livestock production to protect animal welfare by treating uninfected animals, thereby preventing epidemic spread of infectious animal diseases. This provides for a high efficiency of animal production. Antibiotic use is a potential risk for the development of antimicrobial resistant bacteria since these drugs can be found as residues in animal-derived food products [1-4].

Antibiotics used in veterinary and human medicine include mainly: penicillins, cephalosporins, tetracyclines, fluoroquinolones, sulfonamides and macrolides. Penicillins (PENs) have been widely used for more than 30 years and today are one of the most important groups of antibiotics [5].

Residue analysis is closely related to food safety. It establishes whether food is or is not safe for human consumption. Residue analysis is part of the monitoring programs of regulatory agencies. Its aim is to guarantee that concentration levels are below the established maximum residue levels (MRLs). Regulatory demands for the control of chemical contaminants in food have led to an increase in the demand for analytical methods for detecting concentrations below MRLs in food stuffs.

Advances in analytical instrumentation have been invaluable in assisting in each step of the analytical cycle. Different strategies have been developed for controlling the presence of PENs

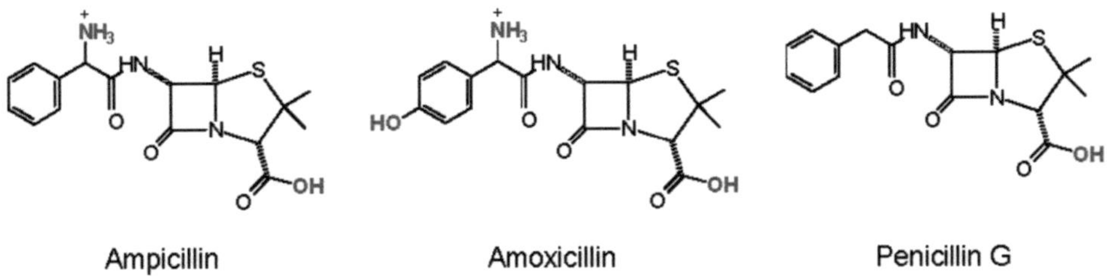

Figure 1. Chemical structures of the main PENs

residue in food samples. These methods vary in reliability, the speed of obtaining results, and cost of the analysis. These methods can be grouped into the following categories [6]:

1. Microbiological approaches based on bacterial growth inhibition
2. Electroanalytical methods
3. Immunochemical techniques
4. Separation techniques

Separation methods, such as gas chromatography and high performance liquid chromatography, are the techniques commonly used to determine the low concentration levels of this antibiotic group. Capillary electrophoresis (CE) is increasingly being used for confirmation purposes. The main drawback of CE is the low sensitivity when UV detection is used, as a consequence of the short optical path length and the extraction of PENs from complex matrices. Several sample treatments are required in most cases to extract and pre-concentrate the analytes. Additionally, during CE analysis some inorganic ions, proteins, and other major compounds can affect the electrophoretic mobility [7].

Recently, a review article focused on PENs analysis by CE can be found in the literature [8]. It describes the importance of high sample volumes in the analysis as an alternative for obtaining lower limits of detection. In this study, we propose the combination of magnetic solid phase extraction and capillary electrophoresis in the analysis of PENs in milk samples.

1.1. Penicillins

Penicillin was discovered by Alexander Fleming in 1928 and was purified and then synthesized by Florey and Chain in 1940. PENs belong to the β-lactam antibiotics. Their structure consist of a thiazolidine ring attached to a β-lactam ring, forming 6 aminopenicillanic acids and a side chain in the 6-position, which determines the stability and the antimicrobial activity of the different derivatives. The mechanism of action of PENs is via the inhibition of bacterial cell death of the offending bacteria due to the faulty production of the vital cell wall components [9].

The most common PENs are ampicillin (AMP), amoxicillin (AMO) and penicillin G (PEN G). Figure 1 shows the acid structure of the main PENs.

The existence of PENs charge molecules suitable for electrophoretic separation is a result of the presence of acid-base functional groups in their structure such as amino, carboxylic acid, and phenol. They can exhibit cationic and zwitterionic forms when an amino group is presented, while the carboxylic and phenol groups contribute to generate anions. The pKa values are: ampicillin (pKa_1=2.6, pKa_2=7.4), amoxicillin (pKa_1=2.6, pKa_2=7.2, pKa_3=9.6), and penicillin (pka_1=2.6). [10]

β-lactams are usually employed in treating mastitis in cows, therefore milk is the most frequently analyzed sample. As a result, the international organizations have defined MRLs for PENs. The European Union directive establishes MRLs of 4 $\mu g\ kg^{-1}$ for the three main PENs [11]. On the other hand, the U.S. Food and Drug Administration established the MRLs for these drugs as follows: amoxicillin and ampicillin, 10 $\mu g\ kg^{-1}$; and penicillin G, 5 $\mu g\ kg^{-1}$ [12]. In order to achieve the MRLs, sample preparation processes including both clean-up and pre-concentration are indispensable during determination of PENs residue in milk samples.

1.2. Sample preparation

PENs exhibit significant binding to proteins contained in milk samples, thus the first step during milk analysis by CE is to deproteinize the sample. Additionally, this procedure helps to prevent emulsions during partition equilibriums. The main strategies used are acidic precipitation or addition of acetonitrile. Methanol should be avoided as a precipitation solvent because it degrades PENs to the corresponding alkyl-α-D-penicilloic acids. [13]. Strong acidic deproteinizong agents are not recommended because of the degradation of PENs. However, good recoveries have been described when trichloroacetic acid was used for the precipitation of milk samples containing PENs concentrations higher than the MRLs [14].

Acetonitrile provided satisfactory results for precipitating proteins and, since the solubility of lipids in the organic solvent is low, most of the lipid fraction was co-precipitated. For the recovery of PENs from milk, a volume ratio of acetonitrile:milk higher than 2:1 was optimal. The use of lower ratios has been described for the analysis of PENs during the first step using the QuEChERS technique [15].

Once the sample is deproteinize, the acetonitrile phase is separated and evaporated. The sample is reconstituted in aqueous phase to be suitable for solid phase extraction. Oasis HLB and C18 cartridges have been applied during sample pre-concentration and clean-up. Acetonitrile was commonly used for the elution of the sample. When an on-line pre-concentration system such as large volume sample stacking is employed, a low conductivity sample is required. The acetonitrile extract is then passed through a polar sorbent (alumina N) and the PENs are eluted with deionized water.

Recently discovered, dispersive techniques (liquid and solid) have shown to improve the contact between the analyte and the extracting phase [16-17]. A clear example is the so-called QuEChERS process, which involves a second clean up step based on dispersive solid phase extraction. During PENs analysis, a mixture of silica C18, primary and secondary amine and acetonitrile were added to an aliquot of the organic extract in order to remove sugars and fatty acids. The extract was centrifuged, evaporated, and reconstituted in a phosphate buffer before its analysis.

There are a few reports related to the analysis of PENs in milk samples by CE. Table 1 shows the conditions used for the preparation of a milk sample for analysis of AMP, AMO and PEN G.

Recovery of the solid phase in dispersive solid phase extraction requires the use of additional procedures such as filtration or centrifugation. The loss of solid during extraction might affect precision and accuracy. In order to minimize the steps involved, magnetic compounds have been incorporated (mainly, Fe_3O_4 or γ-Fe_2O_3) in the solid phase. These magnetic supports can be dispersed and isolated from the sample by the use of an external magnetic field [18-19].

Sample pre-treatment	Separation mode	Limit of detection	Reference
a) 50 ml milk+20 ml trichloroacetic acid (20%) b) SPE (Oasis-HLB), elution with ACN c) Evaporation and reconstitute in 5 ml of H_2O:ACN (1:1)	CZE-UV	0.48-1.09 μg ml^{-1}	[14]
a) 5 ml milk+15 ml ACN b) SPE (Oasis-HLB), elution with ACN c) SPE (Alumina N), elution with H_2O d) Evaporation and reconstitute in 1 ml of H_2O e) On-line concentration	CZE-DAD	4 μg kg^{-1}	[17]
a) 2 mL milk + 4 mL ACN b) The liquid phase was dried and reconstituted in H_2O (1 ml) d) SPE (C18 cartridge), elution with ACN e) Evaporation and reconstitute in H_2O (1 ml) f) On-line sample concentration	CEC-MS	0.05-0.11 μg l^{-1}	[10]
a) 50 ml milk+HCl to reach a pH 3.4-3.8 b) SPE (Oasis HLB), elution with ACN:MeOH (1:1 v:v) c) Evaporation and reconstitute in H_2O (0.25 ml)	MEKC-UV	2.4-3.2 μg kg^{-1} (in water)	[15]
a) 10 g milk+ 8 ml (PBS) + 10 ml (ACN) + QuEChERS mixture b) 4 ml of liquid phase + dispersive SPE (C18, PSA, $MgSO_4$) c) Evaporation and reconstitute in H_2O (0.25 ml)	MEKC-UV	N.R.	[15]

SPE: solid phase extraction, **ACN**: acetonitrile, **MeOH**: methanol, **CZE**; capillary zone electrophoresis, **CEC**: capillary electrophoresis, **MEKC**, micellar electrokinetic chromatography, **UV**: ultraviolet, **DAD**: diode array detection, **MS**: mass spectrometry.

Table 1. Summary of the main PENs analysis in milk samples by CE

Core-shell particles with paramagnetic core and silica shell are the most common particles synthesized. The magnetic solids offer adequate surface area and the possibility of functionalization and paramagnetic properties. Their application as dispersed sorbents in liquid samples is called magnetic solid phase extraction (MSPE) [20]. This technique has demonstrated several advantages such as the decrease in sample treatment time, the decrease in solvent use, and the easy treatment of high volume samples [21-22]. The basic procedure involved in MSPE is represented in Figure 2. Initially, the pre-activated magnetic solid is added to the sample (Fig. 2.a.), the solid is dispersed in the sample and the analytes are retained on the support (Fig. 2.b). Once the extraction is complete, a magnet is placed on the external wall of the sample container to isolate of the solid containing the analytes (Fig. 2.c). Finally, the analytes are eluted from the solid using an appropriate organic solvent (Fig. 2.d).

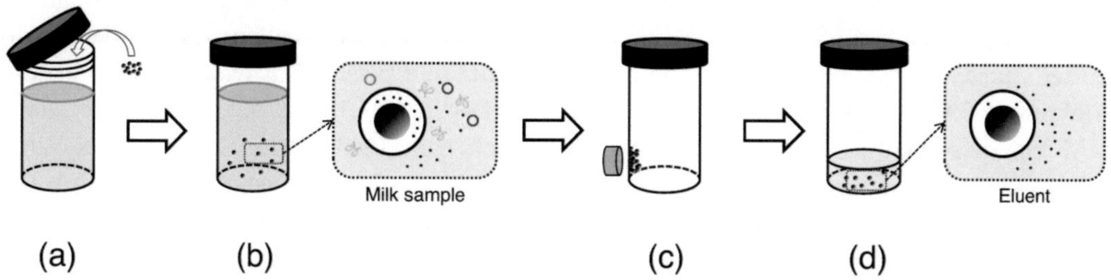

Figure 2. Representation of the magnetic solid phase extraction methodology

MSPE can be coupled with different analytical techniques including gas and liquid chromatography and CE. MSPE-CE has been applied for the selective separation of tetracyclines and fluoroquinolones from milk samples. The possibility of modification has allowed the synthesis of magnetic solid modified with phenyl and phenyl-octyl in order to improve the selectivity during the sample treatment through mixed interaction modes [22-23].

One of the limits of detection obtained with CE-UV is that it requires the use of pre-concentration strategies. On-line concentration by large volume sample stacking has been successfully applied. Taking into account that MSPE allows the use of high volume samples (until 1 liter) during the extraction process, simultaneously with the use of dispersive extraction mode, we propose the analysis of PENs by MSPE-CE-UV.

2. Experimental conditions

2.1. Reagents and dissolutions

All solutions were prepared in deionized water with specific resistivity greater than 18 MΩ cm. PENs solutions were prepared daily by dilution of a 1.0 g l^{-1} standard solution of AMO, AMP, and Pen G (all reagents are analytical grade, Sigma, St. Louis, MO, USA). The background electrolyte or buffer solution used for analysis by MEKC consisted of a mixture of: 20 mM sodium tetraborate and 50 mM sodium dodecyl sulfate (SDS) at pH 8.5. The internal standard used was acid (S)-6-methoxy-α-methyl-2-naphthaleneacetic, which was dissolved in the background electrolyte at a concentration of 5 mg l^{-1}. All the diluted solutions were stored at 4°C for a period no more than seven days.

Milk samples analyzed included 5 of pasteurized milk, 5 of ultra-pasteurized milk and 3 without heat treatment. All the samples were acquired in markets in Pachuca, Hgo, Mexico.

2.2. Synthesis of magnetic support

Magnetite particles (Fe_3O_4) were synthesized as follows: 12 g of $FeSO_4.7H_2O$ was dissolved in 100 ml of water. The mixture was heated with stirring until it reached 60°C. The pH value was adjusted to 10 while airflow was bubbled during the reaction. After 1 hour, the magnetic precipitate was isolated from the solution with a magnet and it was washed with 50 ml portions of deionized water several times. The magnetite obtained previously was added to a flask containing 5.910 g of octyltriethoxysilane (C8-TEOS) and 1.182 g of tetramethoxysilane (TMOS), dissolved in 24.0 mL of a solution which contained: Triton X-100 (2.0%, w/v), cetyltrimethylammonium, methanol (12.5%, v/v), and 200.0 µl of catalyst (NH_3 28 %, v/v). The mixture was heated and refluxed at 120°C for 12 hours with stirring. The obtained magnetic particle was dried at 60°C for 120 hours [21].

2.3. Characterization of the magnetic solid

The structures of the magnetic solid were examined by X-ray diffraction (XRD) patterns using a PHILIPS PW1710 instrument equipped with a Cu anode, automatic divergence slit, and a graphite monochrometer under the following experimental conditions: CuKα radiation, 1.54 Å; generator tension, 40 kV; generator current, 30 mA; intensity ratio ($\alpha 2/\alpha 1$), 0.500; divergence slit, 1°; receiving slit, 0.1; start angle (2θ), 5; end (°2θ), 70. The morphological analysis of the products was performed using a JEOL JSM-820 scanning electron microscope (SEM).

2.4. Sample treatment

The MSPE consisted in the activation of 0.1 g of the magnetic support with 4.0 ml of methanol for 5.0 min in an ultrasonic bath; the solid was isolated magnetically and washed with 5.0 ml of deionized water for 4 min, then the milk sample was added (50,100 and 200 ml), dispersing the magnetic solid for 8.0 min. Once the dispersion was concluded, the magnetic support was magnetically isolated and the milk was decanted and the solid was washed with 20 ml of phosphate buffer solution (pH 7.0, 0.1 M). Finally, the adsorbed analytes were eluted with 5.0 ml of acetonitrile for 4 min and the organic phase was evaporated under an air steam until dried. The residue was reconstituted in the background electrolyte containing the internal standard at 5 mg l^{-1} concentration.

2.5. Procedure of separation by MEKC

Capillary electrophoresis experiments were performed on a Beckman Coulter P/ACE 5500, with a diode array detector. The data obtained was analyzed by the P/ACE MDQ software version 2.3. Separations of the PENs were performed in a fused-silica capillary (57 cm× 50 µm I.D.). The detection wavelength was set a 200 nm and the temperature capillary was thermostated at 30°C during separation. The separation voltage was set at 16 kV. The sample was introduced using a pressure of 0.5 psi.

Initially, the capillary was rinsed with 1.0 M NaOH (20 min), followed by 0.1 M NaOH (20 min), deionized water (15 min), and finally with the background electrolyte (20 min). Each separation in samples was preceded by a rinse with 1.0 M NaOH (2 min), 0.1 M NaOH (2 min), followed by H_2O (5 min), and background electrolyte (5 min).

3. Results and discussion

3.1. Optimization of background electrolyte composition

PENs can be separated in different modes including capillary zone electrophoresis, micellar electrokinetic chromatography, and capillary electrochromatography. Different background electrolytes have been described for PENs analysis in different matrices. The pH value used is in the interval from 3.6 to 9.2 using phosphate or borate and sodium dodecyl sulphate salts [24-28].

Figure 3 shows the predominance zone diagrams for the studied PENs using the pKa values reported for the analytes. The zone marked in the Figure denotes the pH values used for separation of PENs. It can be seen that neutral and anionic forms are presented, then MEKC (using SDS) is the most suitable mode for the separation of PENs.

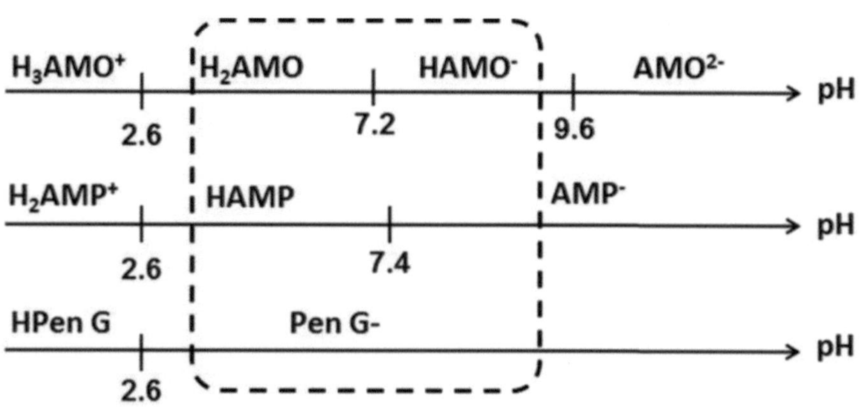

Figure 3. Predominance zone diagrams for PENs

Different factors affecting the separation with the background electrolyte required evaluation. A Taguchi parameter design (TPD) was selected as the optimal method since this provides the necessary information with the minimal experimentation. TPD discriminates between control factors, treating them separately by a means of special design matrices (orthogonal arrays) in which the columns (corresponding to factors) and rows (corresponding to trials) are arranged in a conveniently fixed manner. These matrices indicate the combination of factor settings in each experiment and allows for the simultaneous evaluation of several variables with the minimum number of trials. The results obtained were analyzed statistically to adjust each variable to its optimum [29].

Optimization of the system with TPD involves 5 steps, I) identifying the output variable to optimize, II) identifying and selecting factors that affect the system, III) selecting the appropriate orthogonal array and assigning adequate settings to the chosen factors, IV) analyzing the data and determining the optimum settings, and V) conducting a confirmatory experiment under the optimal conditions obtained [30].

In CE techniques, the effect of the background composition on the separation can be evaluated using the sum of the electrophoretic mobilities (μ_{ion}) of the ions evaluated according the following relation:

$$\mu_{ion}=\frac{L_t L_d}{V}\left[\frac{1}{t_{ion}}-\frac{1}{t_{eof}}\right]$$

Where L_t is the total capillary length (0.57 m), Ld is the capillary length from the injection inlet to the detector (0.32 m), V is the applied voltage (16 kV), and t_{ion} and t_{eof} are the migration times (s) of the analyte and the electroosmotic flow, respectively [14].

The composition (concentration of borate and SDS) and pH value are the variables involved. The selected orthogonal array must have a number of columns equal to or higher than the number of degrees of freedom of the system; thus an $L_9(3^3)$ array was used. The three settings selected for each factor were chosen using MEKC electrolytes reported during PENs analyses. The concentrations of borate and SDS affect the conductivity and the micelle formation. With respect to the pH, the value influences the formation of anionic forms of the analytes. Table 2 shows the settings for each control and noise factor used in the optimization experiments.

Factors	**Level**		
	1	**2**	**3**
[BO_3^-] (mM)	10	20	30
[SDS] (mM)	50	60	70
pH	7.5	8.0	8.5

Table 2. Control factor settings for the optimization experiment

Table 3 shows the factorial design matrix and the results obtained for each trial. All experiments were performed in duplicate in order to calculate the residual error; the total number of experiments was therefore 9 experiments x 2 replicates. Measurements were performed with solutions containing 50.0 mg l^{-1} of each PEN.

http://dx.doi.org/10.5772/59371

Exp	Control factors and levels			$\lvert \Sigma\mu_{ion} \rvert$ *
	$[BO_3^-]$	[SDS]	pH	(10^{-8} $m^2V^{-1}s^{-1}$)
1	1	1	1	3.21
2	1	2	2	3.31
3	1	3	3	3.48
4	2	1	2	4.65
5	2	2	3	3.88
6	2	3	1	4.41
7	3	1	3	4.71
8	3	2	1	4.08
9	3	3	2	2.58

* mean values (n=2)

Table 3. $L_9(3^3)$ orthogonal array and sum of ion mobilities (each value is the mean of two readings).

The results were analyzed statistically to adjust each variable to its optimum with the least variability possible. All calculations were made using ANOVA-TM v2.5 software. Table 4 shows the results for these analyses. The values of the variance ratio (*F*) and the critical variance ratio (3.98 for a 95% confidence level) show that all the factors taken into account were critical ($F_{calculated} > F_{critical}$). The factor with the greatest influence on the response was the borate concentration, which accounted for 55.0% of the total variance of the results, followed by SDS concentration (28.8%). The contribution of the residual error was 0.1%; this indicates the correct selection of experimental parameters.

Figure 4 shows the effects of the control factors on the output variable (sum of ion mobilities), among which the sample volume is the most important. The combination of settings that allowed the highest output peak was $[BO_3^-]$ 20 mM, [SDS] 50 mM and pH 8.5.

Variance source	Degree of freedom	Sum of squares	Variance	Variance ratio (F)[a]	Influence (%)[b]
$[BO_3^-]$	2	102.77	51.39	2569.50	55.0
[SDS]	2	53.68	26.84	1344.50	28.8
pH	2	30.22	15.11	755.50	16.1
Residual	11	0.22	0.02	1.00	0.1
Total	17	186.69			100.0

[a] The critical variance ratio for a 95% confidence level is 3.98 (2,11 d.f.)

[b] Contribution is defined as 100 x (sum of squares/total sum of squares)

Table 4. ANOVA for sum of ion mobilities in Table 3.

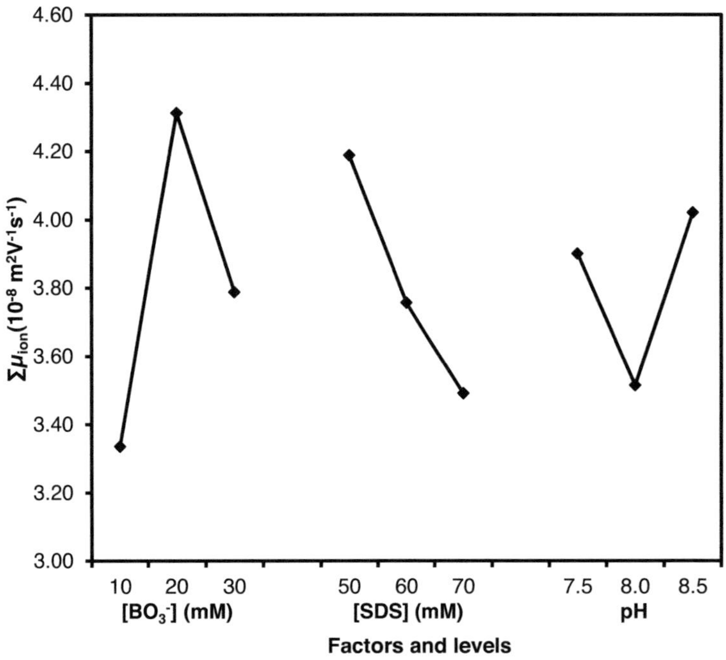

Figure 4. Effect of control factors on the sum of ion mobilities.

3.2. Characterization of magnetic solid

The results of analysis by the XRD are displayed in Fig. 5.a; the diffractogram obtained shows diffraction lines characteristic of magnetite (2θ=30.1°, 35.5°, 43.1°, 53.4°, 57.0° and 62.6°) expressed in the Figure as "m". The wide band observed at 2θ angle from 10° to 25° is consistent with the amorphous phase of silica gel [31]. The micrograph obtained for the magnetic support (Fig. 5.b) shows a spherical morphology for the solid with a particle size around to 1-2 μm.

Figure 6 shows FTIR spectra for the magnetic supports. Magnetite has an inverse spinal type structure with a characteristic vibration M_T-O-M_O at ≈ 600-550 cm^{-1}, where M_T and M_O correspond to the metal occupying tetrahedral and octahedral positions, respectively. On the other hand, the spectra showed a band of intense stretching at ≈3450 cm^{-1}, attributable to the vibration of the silanol group (SiOH). The peaks below 3000 cm^{-1} corresponds to the vibration of-CH_2-. The bending band at ~1630 cm^{-1} corresponds to H_2O occluded in the support. The flexion vibration band at ~1400 cm^{-1} represents the C-H bond of the alkyl groups joined to the silicon. The stretching band at 1100-1000 cm^{-1} is assigned to the Si-O-Si bond and the deformation band at ~888 cm^{-1} belongs to Si-OH bonds [32].

The synthesis protocol used to obtain the magnetic solid is based on the hydrolysis of alkoxysilanes. This process produces solids with silanol (Si-OH) and octyl groups (Si-C_8H_{17}), modifying the retention mechanism [23,33].

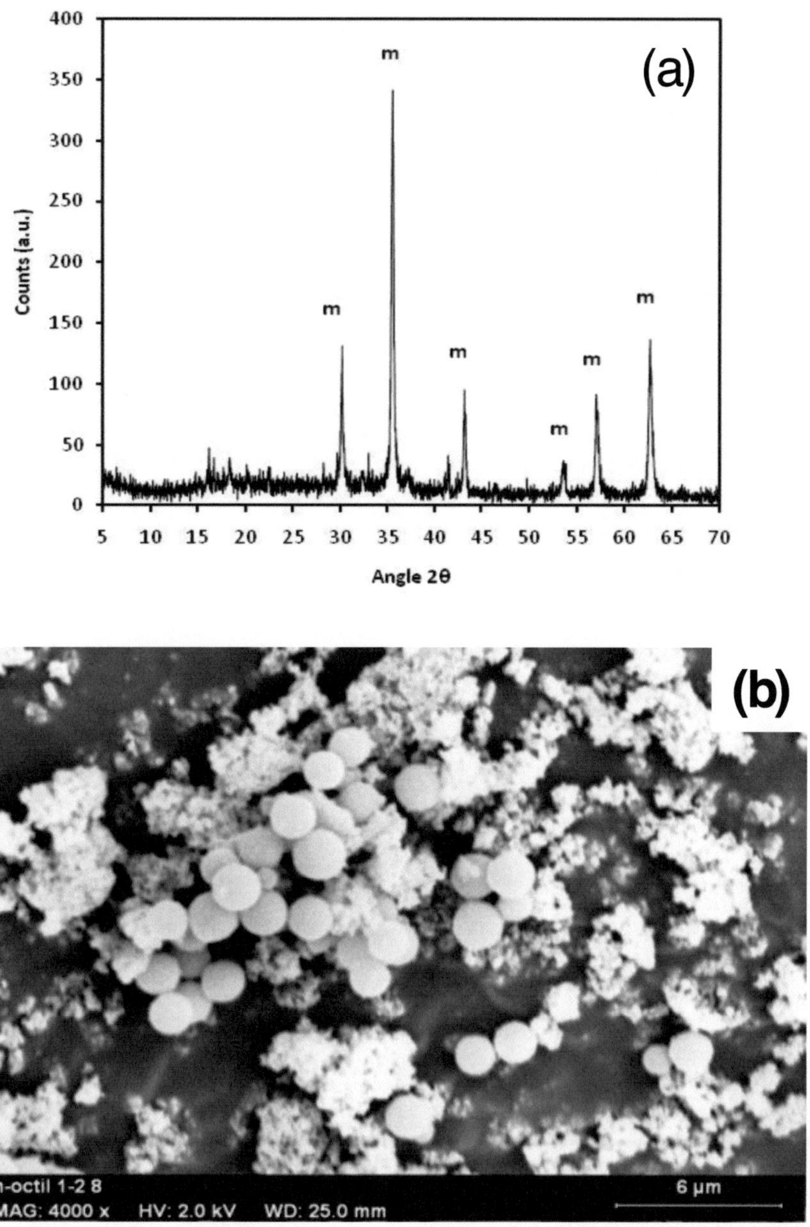

Figure 5. a) Diffractogram of the diffraction lines of magnetite of C_8-TEOS; b) Scanning electron micrograph of the magnetic C_8-TEOS

Because they have a higher affinity for the polar part of the solid, proteins interact with the silanol groups, releasing penicillins, which then are partially retained by the MSPE octyl group.

Thus, the presence of both groups promotes a mix mode retention mechanism. In consequence a single MSPE process can be performed without a protein precipitation and multiple SPE procedures [34].

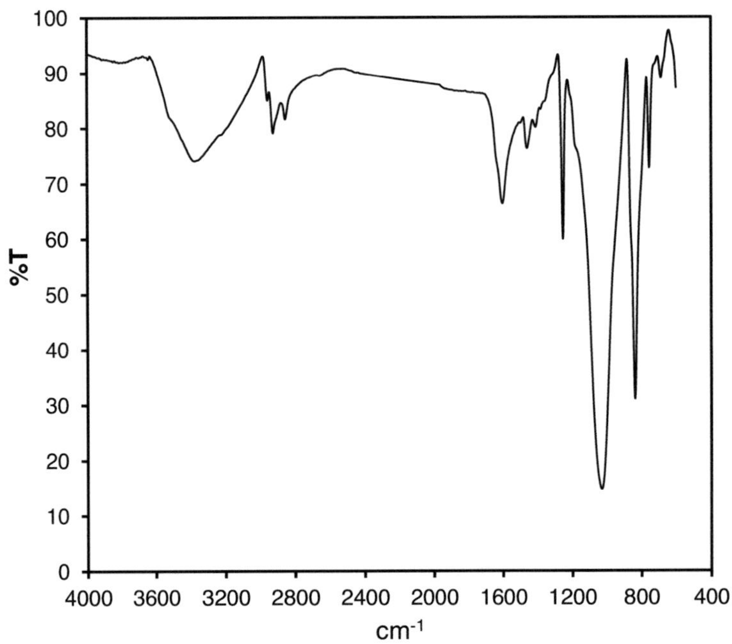

Figure 6. FTIR spectra for magnetic solid (C8) used for MSPE experiments

3.3. Analysis of PENs by MEKC-CE

Quantification of PENs at maximum residual limits involves clean-up and pre-concentration steps. In order to achieve the limits of detection (LOD) required by the international norms, the effect of initial sample volume used on the MSPE methodology was evaluated in the range from 25 to 200 ml. Table 5 shows the LOD obtained; the values decrease with increasing volume. Since a volume of 200 mL yields adequate LODs, it was chosen as the optimal original sample volume.

Sample volume initial (ml)	Limits of detection obtained (µg l^{-1})
25	100.0-120.0
100	24.0-30.0
200	0.20-3.0

Table 5. Limits of detection (µg l^{-1}) reached at different initial sample volumes.

Once the appropriate initial sample volume for analysis is established, a series of penicillin standard solutions with concentrations ranging from 10-70 μg l^{-1} were analyzed. Table 6 shows the evaluation of analytical performance of the MEKC methodology. Precision of the method (reproducibility and repeatability) was expressed as relative standard deviation (%RSD) calculated from three assays using a concentration of 10 μg l^{-1} penicillin. A maximum %RSD value of 20% is required during ultratrace analysis, thus the PENs determination with the proposed method is precise [1,35].

Parameter	AMO	AMP	PEN G
Slope confidence interval $b_1 \pm t\ s(b_1)$	0.267±0.012	0.342±0.007	0.709±0.001
Intercept confidence interval $b_0 \pm t\ s(b_0)$	0.027±0.002	0.023±0.001	0.032±0.002
Correlation coefficient, r^2	0.999	0.999	0.999
Limit of detection (μg l^{-1})	2.7	1.7	0.2
Limit of quantification (μg l^{-1})	8.1	5.1	0.6
Linear range (μg l^{-1})	2.7-60	1.7-60	0.2-60
Repeatability (%RSD, 10 μg l^{-1}, n=3)	6.8	1.2	5.1
Reproducibility (%RSD, 10 μg l^{-1},n=3)	13.0	9.9	11.6

Table 6. Analytical parameters obtained of analysis of penicillins in milk.

The accuracy was evaluated by recovery assays through spiked milk samples with the penicillins studied in a range of 10-30 μg l^{-1}. Table 7 shows the recovery percentages obtained from the assay. The results generated by the proposed methodology are accurate according to the determined recovery percentages. The recovery must be in the range of 80-120% during ultratrace analysis to be considered an acceptable parameter [1].

Concentration (μg kg^{-1})	% Recovery		
	AMO	AMP	PEN G
10	87	97	106
20	97	80	83

Table 7. Recovery values obtained of analysis of spiked milk samples.

The developed method was applied to the determination of penicillins in 15 commercial milk samples. Three replicate determinations were carried out on each sample. Overall, two samples of ultrapasteurized milk were found positive for the presence of residues of penicillins, one with AMP and other with AMO. The concentrations found were 35±1 and 34±1 μg l^{-1} for AMP and AMO, respectively. The concentration found is higher than the MRLs. Figure 7 shows the electropherograms obtained during the analysis of the spiked blank sample (Fig. 7a), commercial milk sample positive to ampicillin (Fig. 7b), and the blank sample (Fig. 7c). These results are congruent with the antibiotic resistance found for *Escherichia coli* isolated from Food Stuffs in Hidalgo [36].

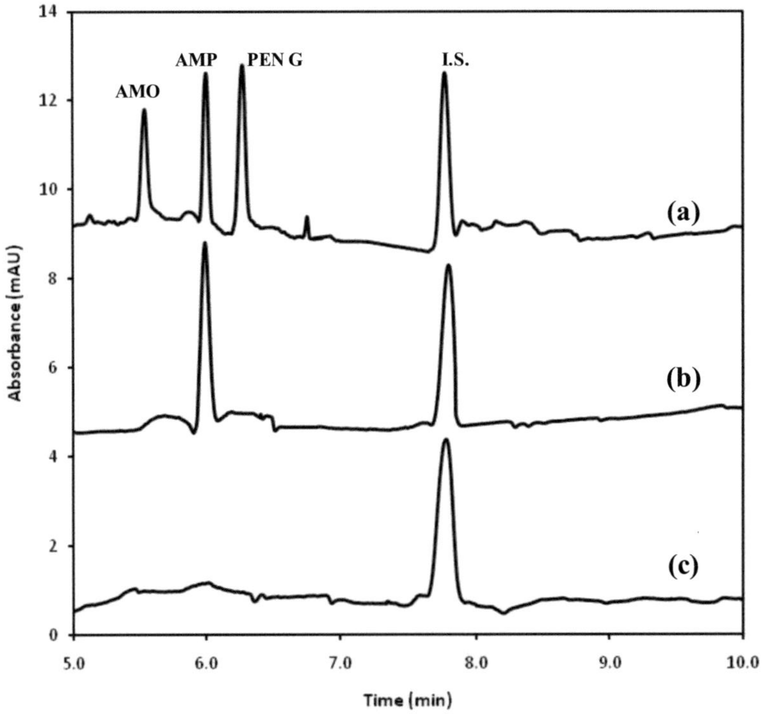

Figure 7. Electropherograms obtained at the optimized conditions in the analysis of: a) PENs spiked blank sample, b) commercial milk sample analyzed by MSPE and c) milk blank sample

The mixed mechanism offered by the solid supports contributes to a better clean-up of the sample, as can be seen in the electropherogram. Although the re-usability of the solid extracting phase is not possible due to the adsorption of other components of the sample, the proposed sample treatment has advantages such the use of fewer organic solvents and the time required for sample treatment is faster than SPE methodologies described.

4. Conclusions

In this study, an octyl-silica based magnetic support was synthesized and successfully applied for the separation of penicillin antibiotics from milk samples. The penicillins were isolated and the matrix interferences were eliminated using this support. The extraction technique is a

http://dx.doi.org/10.5772/59371

robust preconcentration technique that can be coupled successfully with MEKC. The methodology described (MSPE-MEKC) is faster than classical sample preparation procedures such as solid phase extraction, with a minimum sample handling, less solvent consumption, and is a promising methodology for routine milk sample analysis. Moreover, the linear range covers the MRL for PEN and the recovery studies are acceptable regarding to the range for ultratrace analysis.

In the analysis of commercial products, some samples contain residues of penicillins at higher levels than the MRLs. These results are consistent with the antibiotic resistance found for *Escherichia coli isolated* from Food Stuffs in Hidalgo, which shows the congruence between MSPE-MEKC and other microbiological studies.

Acknowledgements

The authors wish to thank the Consellería de Cultura, Educación e Ordenacion Universitaria, Xunta de Galicia (project EM 2012/153) and CONACYT (project INFR-2014-227999) for financial support. J.A. Rodríguez, I.S. Ibarra, C.A. Galán-Vidal, and M.E. Páez-Hernández gratefully thank the SNI for the distinction of their membership.

Author details

J.A. Rodríguez[1], E. Martínez[1], I.S. Ibarra[1], J.M. Miranda[2], C.A. Galán-Vidal[1] and M.E. Páez-Hernández[1*]

*Address all correspondence to: paezh@uaeh.edu.mx

1 Área Académica de Química, Universidad Autónoma del Estado de Hidalgo, Pachuca, Hgo, Mexico

2 Dpto. Química Analítica, Nutrición y Bromatología, Facultad de Veterinaria, Universidad de Santiago de Compostela. Lugo, Spain

References

[1] Botsoglou, N.A., Dimitrios J.F. Drug residues in foods. Marcel Dekker; 2000.

[2] Stolker, A.A.M., Brinkman, U.A. Analytical strategies for residue analysis of veterinary drugs and growth-promoting agents in food-producing animals-a review. Journal of Chromatography A 2005; 1067(1) 15-53.

[3] Miranda, J.M., Mondragon, A.C., Martinez, B., Guarddon, M., Rodriguez, J.A. Prevalence and antimicrobial resistance patterns of Salmonella from different raw foods in Mexico. Journal of Food Protection 2009; 72(5) 966-971.

[4] McEwen, S.A., Fedorka-Cray, P.J. Antimicrobial Use and Resistance in Animals. Clinical Infectious Diseases 2002; 34 (Supplement 3) S93-S106.

[5] Lara, F. J., del Olmo-Iruela, M., Cruces-Blanco, C., Quesada-Molina, C., García-Campaña, A. M. Advances in the determination of β-lactam antibiotics by liquid chromatography. TrAC Trends in Analytical Chemistry 2012; 38, 52-66.

[6] Kantiani, L., Farré, M., Barceló, D., Barceló, D. (2009). Analytical methodologies for the detection of β-lactam antibiotics in milk and feed samples. TrAC Trends in Analytical Chemistry 2009;28(6) 729-744.

[7] Pérez-Fernández, V., Domínguez-Vega, E., Crego, A. L., García, M. A., Marina, M. L. Recent advances in the analysis of antibiotics by CE and CEC. Electrophoresis 2012; 33(1) 127-146.

[8] Piñero, M. Y., Bauza, R., Arce, L., Valcárcel, M. Determination of penicillins in milk of animal origin by capillary electrophoresis: Is sample treatment the bottleneck for routine laboratories?. Talanta 2014; 119, 75-82.

[9] Samanidou, V. F., Nisyriou, S. A., Papadoyannis, I. N. Residue analysis of penicillins in food products of animal origin by HPLC: A review. Journal of Liquid Chromatography & Related Technologies 2007; 30(9-10)1145-1204.

[10] Liu, W. L., Wu, C. Y., Li, Y. T., Huang, H. Y. Penicillin analyses by capillary electrochromatography-mass spectrometry with different charged poly (stearyl methacrylate–divinylbenzene) monoliths as stationary phases. Talanta 2012; 101, 71-77.

[11] European Commission, Commission Regulation (EC) No 37/2010 of 22 December 2009. Official Journal of the European Union 2010;L15, 1-75.

[12] Nollet, L. M., Toldrá, F. Food analysis by HPLC. CRC Press;2010.

[13] Agarwal V.K. Analysis of antibiotic/drug residues in food products of animal origin. Plenum Press;1992.

[14] Santos, S. M., Henriques, M., Duarte, A. C., Esteves, V. I. Development and application of a capillary electrophoresis based method for the simultaneous screening of six antibiotics in spiked milk samples. Talanta 2007;71(2) 731-737.

[15] Piñero, M. I., Jurado, N., Bauza, R., Arce, L., Valcárcel, M. Evaluation of different strategies to extract and preconcentrate penicillins present in milk prior determination by capillary electrophoresis. Revista Científica 2014;24(1)55-63.

[16] Diez, C., Traag, W. A., Zommer, P., Marinero, P., Atienza, J. Comparison of an acetonitrile extraction/partitioning and "dispersive solid-phase extraction" method with

classical multi-residue methods for the extraction of herbicide residues in barley samples. Journal of Chromatography A 2006;1131(1)11-23.

[17] Bailón-Pérez, M. I., García-Campaña, A. M., Cruces-Blanco, C., del Olmo Iruela, M. Large-volume sample stacking for the analysis of seven β-lactam antibiotics in milk samples of different origins by CZE. Electrophoresis 2007;28(22)4082-4090.

[18] Aguilar-Arteaga, K., Rodriguez, J. A., Barrado, E. Magnetic solids in analytical chemistry: A review. Analytica Chimica Acta 2010;674(2)157-165.

[19] Safarik, I., Horska, K., Pospiskova, K., Safarikova, M. Magnetic techniques for the detection and determination of xenobiotics and cells in water. Analytical and Bioanalytical Chemistry 2012;404(4)1257-1273.

[20] Safarikova, M., Safarik, I. Magnetic solid-phase extraction. Journal of Magnetism and Magnetic Materials 1999;194(1)108-112.

[21] Aguilar-Arteaga, K., Rodriguez, J. A., Miranda, J. M., Medina, J., Barrado, E. Determination of non-steroidal anti-inflammatory drugs in wastewaters by magnetic matrix solid phase dispersion–HPLC. Talanta 2010;80(3)1152-1157.

[22] Ibarra, I. S., Rodriguez, J. A., Miranda, J. M., Vega, M., Barrado, E. Magnetic solid phase extraction based on phenyl silica adsorbent for the determination of tetracyclines in milk samples by capillary electrophoresis. Journal of Chromatography A 2011;1218(16)2196-2202.

[23] Ibarra, I.S., Rodriguez, J.A., Páez-Hernández, M.E., Santos, E.M., Miranda, J.M. Determination of quinolones in milk samples using a combination of magnetic solid-phase extraction and capillary electrophoresis. Electrophoresis 2012;33(13)2041-2048.

[24] Perez, M. I., Rodríguez, L. C., Cruces-Blanco, C. Analysis of different β-lactams antibiotics in pharmaceutical preparations using micellar electrokinetic capillary chromatography. Journal of Pharmaceutical and Biomedical Analysis 2007;43(2)746-752.

[25] Bailón-Pérez, M. I., García-Campaña, A. M., del Olmo Iruela, M., Cruces-Blanco, C., Gracia, L. G. Multiresidue determination of penicillins in environmental waters and chicken muscle samples by means of capillary electrophoresis-tandem mass spectrometry. Electrophoresis 2009;30(10)1708-1717.

[26] Nozal, L., Arce, L., Ríos, A., Valcárcel, M. Development of a screening method for analytical control of antibiotic residues by micellar electrokinetic capillary chromatography. Analytica Chimica Acta 2004;523(1)21-28.

[27] Puig, P., Borrull, F., Aguilar, C., Calull, M. Sample stacking for the analysis of penicillins by microemulsion electrokinetic capillary chromatography. Journal of Chromatography B 2006;831(1)196-204.

[28] Thomas, A., Ukpoma, O. K., Inman, J. A., Kaul, A. K., Beeson, J. H., Roberts, K. P. Quantification of penicillin G during labor and delivery by capillary electrophoresis. Journal of Biochemical and Biophysical Methods 2008;70(6), 992-998.

[29] Rodriguez, J. A., Espinosa, J., Aguilar-Arteaga, K., Ibarra, I. S., Miranda, J. M. Determination of tetracyclines in milk samples by magnetic solid phase extraction flow injection analysis. Microchimica Acta 2010;171(3-4) 407-413.

[30] Hernandez, P., Rodriguez, J. A., Galan, C. A., Castrillejo, Y., Barrado, E. Amperometric flow system for blood glucose determination using an immobilized enzyme magnetic reactor. Biosensors and Bioelectronics 2013;41 244-248.

[31] Zou, W., Geng, H., Lin, M., Xiong, X. Facile one-pot preparation of superparamagnetic chitosan sphere and its derived hollow sphere. Journal of Applied Polymer Science 2012;123(6)3587-3594.

[32] Barrado, E., Rodríguez, J. A., Prieto, F., Medina, J. Characterization of iron oxides embedded in silica gel obtained by two different methods. Journal of Non-Crystalline Solids 2005;351(10)906-914.

[33] Benito-Peña, E., Partal-Rodera, A.I., León-González, M.E., Moreno-Bondi, M.C. Evaluation of mixed mode solid phase extraction cartridges for the preconcentration of beta-lactam antibiotics in wastewater using liquid chromatography with UV-DAD detection. Analytica Chimica Acta 2006;556(2)415-422.

[34] Hage, D. S., Tweed, S. A. Recent advances in chromatographic and electrophoretic methods for the study of drug-protein interactions. Journal of Chromatography B: Biomedical Sciences and Applications 1997;699(1), 499-525.

[35] Currie, L. A. (1995). Nomenclature in evaluation of analytical methods including detection and quantification capabilities. Pure and Applied Chemistry 1995;67,1699-1723

[36] Miranda, J. M., Mondragon, A., Rodriguez, J. A., Guarddon, M., Nebot, C. G., Galán-Vidal, C. A., & Coronel-Olivares, C. Presence and antimicrobial resistance of Escherichia coli isolated from foodstuffs in Hidalgo State (Mexico). CyTA–Journal of Food 2010;8(1)15-21.

Chapter 4

Capillary Electrophoresis as Useful Tool in Analysis of *Fagus sylvatica* L. Population Genetic Dynamics

Justyna A. Nowakowska and
Malgorzata K. Sulkowska

Additional information is available at the end of the chapter

http://dx.doi.org/10.5772/59197

1. Introduction

1.1. What kind of tool is Capillary Electrophoresis (CE)

The Capillary Electrophoresis (CE) is one of the method widely used in modern molecular genetics, applied for fast and efficient DNA fragment separation in the sieving polymer in the electric field [1]. The use of this method has increased dramatically over the last fifteen years, due to high precision of small nucleic acid separation (even to the singe nucleotide level) of the available material analysed in the field of analytical chemistry, physical chemistry, biochemistry, and biotechnology. Essentially, the CE technique has been used for genome sequencing projects, e.g. Human Genome Project [2] or many others assignments (for animals, plants, bacteria and fungi), published in NCBI database (www.ncbi.org).

Progress in the CE area in recent years relied more on increasing the number of samples analyzed at the same time, as well as the development of new gels (which allow multiple separation in the same capillary filling), and "chemistry" (a mixture of buffers, substrates, and the so-called polymerase enhancers) for analyzing a sequence of one fragment nearly 1000 base-pairs. Nowadays, the CE technique applied for sequencing analysis is overcome by pyrosequencing method relying on the luminometric detection of pyrophosphate that is released during primer-directed DNA polymerase catalyzed nucleotide incorporation [3].

The CE technique is commonly performed in automated sequencers, i.e. CEQ™ 8000 Genetic Analysis System (Beckman Coulter, Fullerton, CA) composed by two main components: hardware (apparatus) and the CEQ System software. The named model is equipped by 8 capillary system which ensure 8 sample analysis in the same time.

Another apparatus used for CE, especially recommended in case of DNA sequencing, is e.g. 3500 Genetic Analyzer (Life Technologies) which includes the capillary electrophoresis instrument with the workstation and the 3500 Data Collection Software for instrument control, data collection, quality control and autoanalysis of sample files for basecalling and fragment sizing. Auto-analysis can also be performed in this model thanks to the GeneMapper® or GeneMapper® ID-X Softwares.

Both sequencers cited above are equipped with 8 capillaries, but 16-or even more capillary system are available [4]. All type of automated sequencers require appropriate chemistry used for the given goal (genotyping or sequencing), comprising polymer (separation gel), chemical buffers and washing solutions.

The current chapter describes application of DNA based genotyping of *Fagus sylvatica* population by capillary electrophoresis performed in CEQ™ 8000 (Beckman Coulter) sequencer.

1.2. Advantages of Capillary Electrophoresis (CE) and weak points

The CE method is widely used in modern molecular biology science in assessment of gene or allele presence (genotyping) in a given DNA sample, as well as in determining of the DNA nucleotide composition (sequencing) of the studied gene.

In general, the CE technique has several advantages:

- Good tool to support conservation and management of forest trees genetic resources, in order to:
    - Characterize the genetic structure of the forest tree stands
    - Assess the initial gene pool of the population
    - Detect the selection processes and to maintain high level of the natural diversity of forest stands
    - Reflect the history of the stand in relation to the post-glacial migration refugia in Europe and in the world (phylogeny study)
- Provide genetic characteristics of different forest tree species reproductive material in:
    - Mother and progeny stands gene flow analysis
    - Seed orchards and progeny plantation mating system
    - Assignment of populations selected for preservation in gene banks or in situ or ex situ measures
- Solving of problems from seed stand management point of view:
    - Clonal/pedigree identification/selection processes
    - Pollen contamination especially important in management of artificial tree stands like forest seed plantations

  - Patterns of gene flow and mating system in natural and artificial stands
- Tracing of forest tree species with DNA markers as a support in the combat with illegal logging:
  - Thanks to the DNA profiles established on a basis of minimum 4 microsatellite nuclear DNA loci, and at least one cytoplasmic (mitochondrial or chloroplast) DNA marker
  - Strong proof to support the decision taken by several District Law Courts, as far as the identification of wood samples is proved with a high probability (approximately 98–99%)
  - Consistent with the assumptions of the European Parliament Directive on Timber Regulation (EUTR), which came into effect in 2013 to stop the circulation of illegally logged wood in the European Union.

Major weak points of the CE method are derived from:

- Can be avoid by appropriate programme i.e. GeneMapper® or GeneMapper® ID-X Softwares which analyse, process and report the date basing on bank of dataset
- Addition of at least 10% formamide improves the denaturing capacity of the sieving matrix and reduces the compressions
- Can be avoid by appropriate machine running respecting all routine maintenance and installation rules according to the manufacturer.

1.3. Methodical problems

Two main errors may occur during the genotyping procedure performed with CE technique. First of all, some alleles are not identified as peaks in chromatograms (so called null alleles), influencing the general allele distribution in the population. Appropriate software, e.g. GenALEx [5] or Micro-Checker (http://www.microchecker.hull.ac.uk) may help in calculation of the probability of the null allele occurrence in the studied group of trees.

Secondly, the homplasy phenomenon may also occur during genotyping. This term is applied to the DNA fragments of the same size (in base-pairs, i.e. 324 bp) deriving from different microsatellite loci e.g. mcf-5 and mcf-11 in the case of *Fagus sylvatica* species. Such errors are avoided by different fluorochrome labelling of the primers during PCR prior to the CE detection in automated sequencer.

In sequencing data obtained from CE technique, the sample contamination by the DNA molecules from the other species, or errors in trimming of the 5′ and 3′ ends of the coding regions. The trimming errors are estimated to be very low, corresponding to the rate of 0.07, 0.06, 0.05, 0.03, 0.01 with medium default value of 5%.

Good quality of DNA sequences and fragments are obtained thanks to cautious application of the user guide advice provided by the sequencer manufacturer, e.g. CEQ™ 8000 Genetic Analysis System User's Guide (www.beckmancoulter.com/wsrportal/wsr/index.htm) or Applied Biosystems 3500/3500xL Genetic Analyzer User Guide: http://tools.lifetechnologies.com/content/sfs/manuals/cms_069856.pdf.

All those troubleshoots occurring within genotyping and sequencing procedure can be overcome by use of the control samples (individual of reference, with known DNA structure) analysed in the same run. The other possibility of solving problems related with CE analysis is repetition of the experiment from the beginning with new reagents and new PCR reactions (see Methods below).

2. Methods

2.1. The idea of Capillary Electrophoresis (CE) electrophoresis

The assessment of the genetic structure (both for genotyping and sequencing) rely on five basic steps of treatment applied to the plant material collected from the field.

Applied procedure of DNA analysis in CE consists on the following steps:

- Total genomic DNA isolation from the plant tissues (homogenization and extraction of nucleic acid molecules)
- Specific loci amplification via PCR technique using labeled primers
- Separation of the amplified fragments via CE in automated sequencer
- Allele scoring and / or sequence data processing
- Computing of the obtained data base with utilization of proper software.

Concerning the DNA isolation, there are several techniques of extraction, based on the lysis of cell walls to facilitate proper isolation of nucleic acids from the plant, fungal and animal tissues [6]. Good performance of extraction is guaranteed by kits ready for DNA isolation, e.g. DNeasy Plant Mini Kit (QIAGEN®) or NucleoSpin® Plant II (Machery-Nagel).

The first step in obtaining DNA molecules from plant tissues, is initial mechanical homogenization or grinding the material in liquid nitrogen. Liquid nitrogen damages mechanically cell walls, which allows easier access to DNA molecules, at the same time as their stability at low temperature –176°C, is maintained. Efficiency of DNA extraction is then analyzed by spectrophotometry (Fig. 1) or by electrophoresis in agarose gel, followed by staining with ethidium bromide (50 mg ml^{-1}). During sequencing, template impurity due to presence anions coming from unpurified PCR product mix or the presence of another DNA sequence can significantly influence the separation performance of the capillary. At the end, obtained DNA molecules can be stored in stabilizing buffer (pH 7.0) for a long term (even years) at –75°C.

The polymerase chain reaction (PCR) proceeds in DNA multiplication in a thermocycler programmed for multiple (average of 30 to 40) cycles (Fig. 2). Most of DNA techniques are based on amplification of the genomic DNA fragments thanks to the thermostable enzyme *Taq* polymerase deriving from thermophilic bacteria *Thermus aquaticus*.

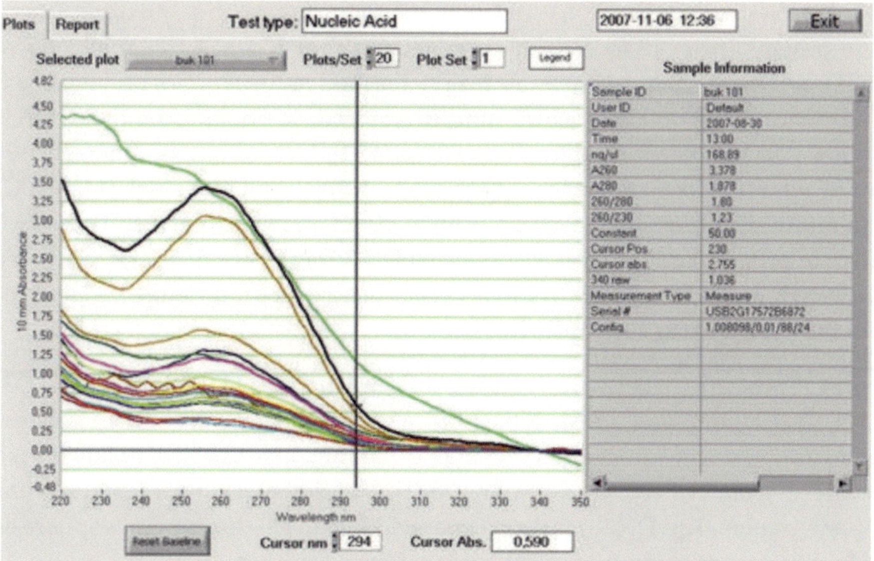

Figure 1. Spectrophotometrical assessment of the DNA extracts from the beech leaf samples in the spectrophotometer NanoDrop® ND-1000 (TK–Biotech, USA). All the DNA extracted from the plant tissues are examined for their purity and evaluation of the DNA concentration before the amplification of microsatellite loci in PCR reaction.

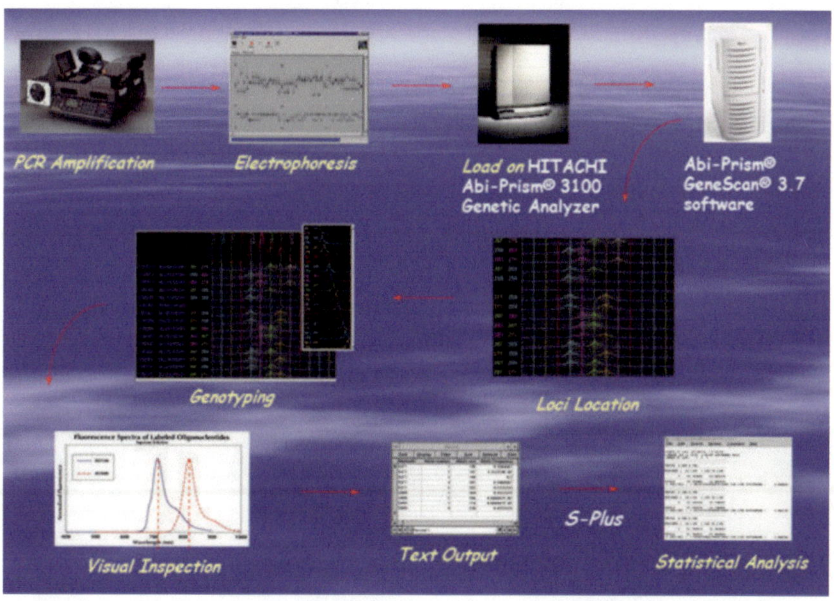

Figure 2. General scheme of the microsatellite loci analysis using the CE in automated sequencer. After PCR amplification followed by gel-electrophoresis, the DNA samples are loaded to the sequencer (e.g. HITACHI Abi-Prism 3100 Genetic Analyzer) for the CE running. After this, the separated DNA fragments are examined in software for precise loci location and genotyping comprising the visual inspection of the peaks (alleles), numeric data collection (list of alleles) and final statistical analysis.

The PCR reaction involves the DNA-matrix in the following reaction mixture: labeled oligonucleotide primers (from 10 to 24 base pairs length), four types of free nucleotides (dATP, dGTP, dCTP and dTTP), magnesium ions (Mg^{2+}), reaction buffer and *Taq* polymerase. The first stage of amplification takes about a few minutes and leads to double-stranded DNA template denaturation at 94°C. Then the stage of annealing at 32– 42°C comes, lasting from 30 sec. to a few minutes, when the formation of complementary DNA strand to the matrix (at 72°C) occurs. The temperature and duration of each stage depends on many factors, mainly on the G/C and A/T content in primers and the size of duplicated DNA fragments. Efficiency and precision of PCR is very high, and theoretically it allows reproducing the output-template DNA molecules present in the extract up to 109 copies [6].

Then, the PCR products, labeled with different fluorochromes (e.g. WellRed D2, D3 and D4 for CEQ™ 8000 model; and FAM, JOE, ROX and TAMRA for Abi-Prism sequencer) are subjected to the CE run in automated genetic analyzer (Fig. 2 and 3). Several labeling strategies have been developed. With some exceptions, the DNA primer labeling is generally done at the 5′ end of one primer. The DNA markers separated in the gel during electrophoresis can be detected by florescence of specific nucleotides in labeled primers.

During electrophoresis in polyacrylamide gel, negatively charged DNA molecules migrate in the direction to a voltage positive electrode of the gel at a speed proportional to their size and molecular weight. After completion of electrophoresis on capillary, the DNA fragments are detected by laser thanks to the appropriate fluorochrome labeling of the primers in PCR reaction (Fig. 3). The separated DNA fragments are then analyzed using computer software.

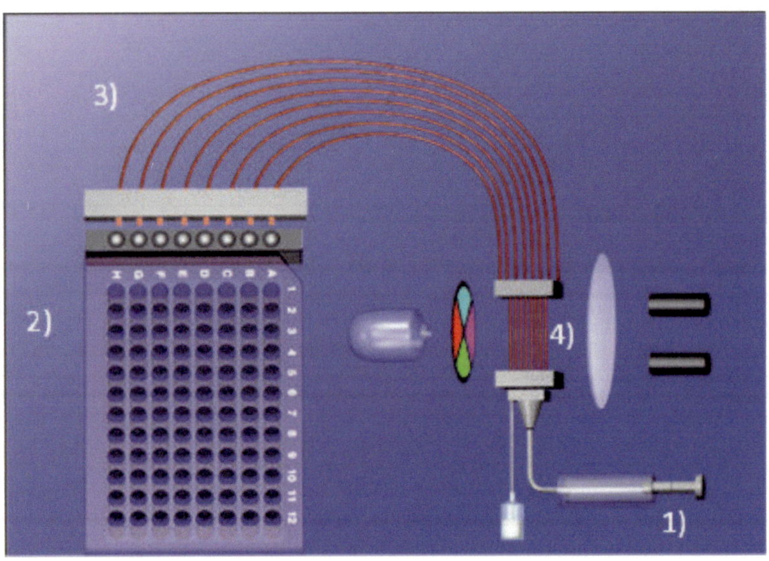

Figure 3. General scheme of the CE principle in the CEQ™ 8000 (Beckman Coulter) sequencer. 1) Gel cartridge – containing the polymer used for separation, 2) 96-well plate with samples, 3) Eight capillaries in which the samples are automatically loaded and the CE is performed, 4) Optical system of windows for the peak detection.

2.2. Genetic differentiation evaluation process

The most crucial step in allele scoring is the exact allele-size determination. For this, a general scheme can be applied, helping to avoid the erroneous allele listing for examined population (Fig. 4). The most tricky allele assignment occurs into 2-base-pair repetition in the SSR fragment, especially when two adjacent alleles for the same locus differ only by 2-bp length (Fig. 4C). Good quality laboratory manipulation and broad experience of the scientist easily overcome such a discrepancy and depict the heterozygous loci in an individual. Otherwise, the false result would lead to the excess of homozygotes in the examined group of trees.

The nuclear microsatellite DNA sequences have so far been considered as the most informative markers, and have been used in the genetic diversity studies of many organisms. They are formed by short repeats of 1-6 base-pairs, so called Short Sequence Repeats (SSR or SSRs), and constitute the most powerful tool in modern population genetics and forensic studies.

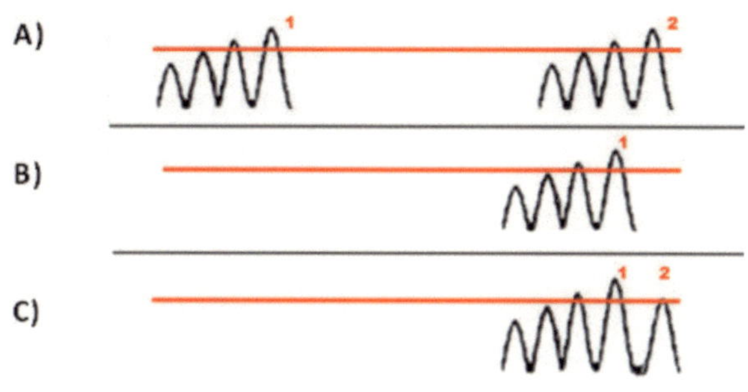

Figure 4. Microsatellite allele scoring, examined according to the appropriate threshold (red line) for peaks obtained from the CE performed in automated sequencer. The biggest peak (i.e. sharing the biggest peak-area) represent the correct fragment size following the rule: A) Well separated Heterozygote, B) Homozygote, C) 2 base-pair separated heterozygote.

The advantages of microsatellite sequences are numerous: they are uniformly distributed over entire genome, are present in high proportion in forest tree species, form discrete loci and co-dominant alleles. The observed mutation rates for the SSR markers vary from 10^{-3} to 10^{-6}. The SSR fragments obtained after the CE technique applied to the European beech populations in Poland illustrate the precision of the detection of different alleles in four nuclear SSR loci investigated (Fig. 5)

The sequencing methodology based on CE relies on nucleotide extension product growing from 5′ to 3′ direction by forming a phosphodiester bridge between the 3′-hydroxyl group at the growing end of the primer and the 5′phosphate group of the incoming deoxynucleotide. The principle of DNA replication follows Sanger dideoxy sequencing procedure [7]. The DNA sequence is copied with high fidelity at each base on the DNA template, as far as the DNA polymerase incorporates only one complementary nucleotide. The resulting nucleotide alignment is registered in chromatogramme and four letter code corresponding to the studied gene fragment (Fig. 6)

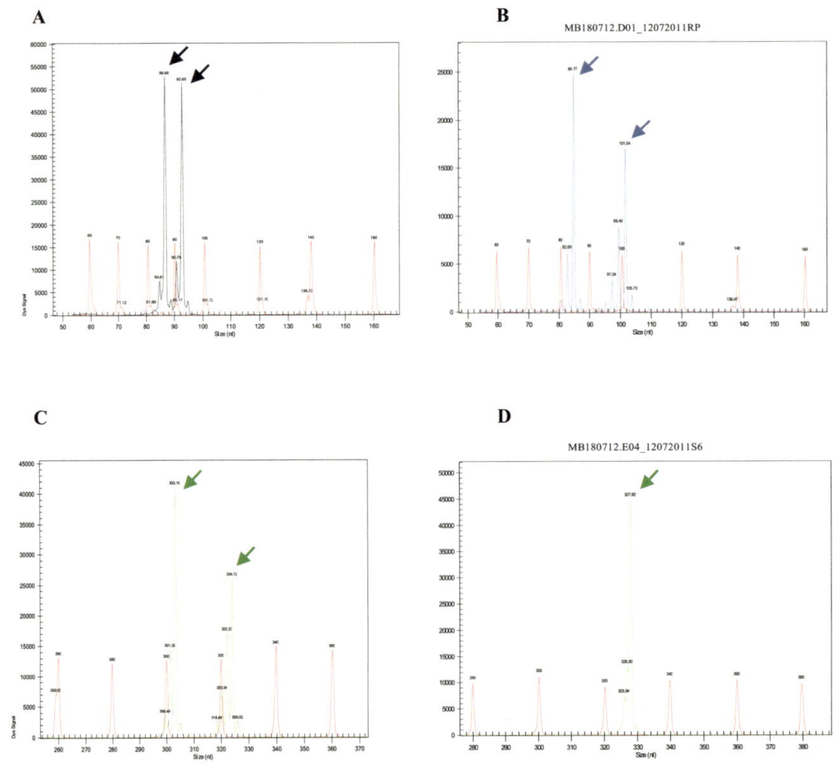

Figure 5. Example of microsatellite nuclear DNA analysis in *Fagus sylvatica* genotyping in population from Bieszczadzki National Park in Poland: two alleles 87 and 93 base-pairs in locus FS1-03 (**A**), two alleles 85 and 101 bp in locus FS1-25 (**B**), two alleles 303 and 324 bp in locus mcf-5 (**C**), one allele 328 bp in locus mcf-11 (**D**). Obtained from DNA capillary electrophoresis after Beckman Coulter® software CEQ™ 8000 Genetic Analysis System v 9.0 (Fullerton, USA).

2.2.1. Analysed parameters to describe population genetic variation and differentiation

The genetic diversity is defined as the probability of occurrence of the identical genotype among randomly chosen trees in a forest stand. The picture of electrophoretic separation of DNA fragments is converted to numerical data, using software such as CEQ System software (Beckman Coulter sequencer) or GeneMapper® or GeneMapper® ID-X Softwares (Abi-Prism sequencer).

The allele size obtained from sequencing data can be checked with the use of S-Plus software version 3.4 release 1 for SPARC (Statistical Sciences, Math Soft Inc., Seattle, WA).

In fact, the population genetic variation and differentiation are based on the heterozygosity parameter (1) calculated as values from zero (no heterozygosity) up to nearly 1.0 (when we observe a large number of almost equally frequent alleles). Instead of average number of alleles per locus more precise measure of effective number of alleles per locus (n_e) – Crow & Kimura [8] can be used (2).

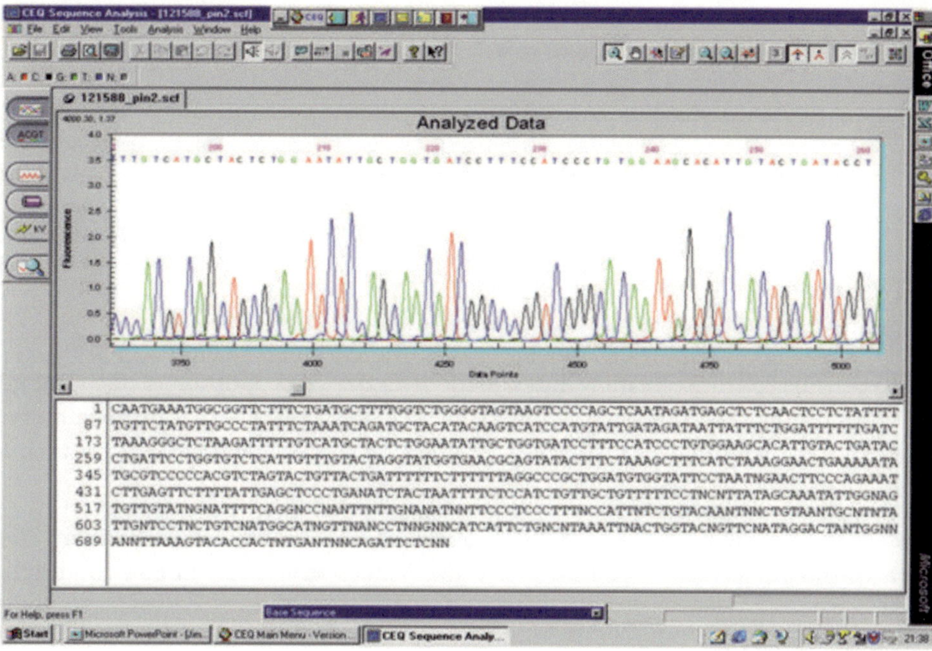

Figure 6. General scheme of capillary electrophoresis data processing for gene sequencing in automated sequencer CEQ™ 8000 (Beckman Coulter). All single nucleotide acids are visible thanks to the different fluorochrome labeling, Adenine (A) in red, Thymine (T) in blue, Guanine (G) in green and Cytosine (C) in black.

$$He = 1 - \sum_{i=1}^{n} pi^2 \quad (1)$$

$$n_e = \frac{1}{\sum_{i=1}^{n} pi^2} \quad (2)$$

p_i – frequency of n allele occurrence in population

Heterozygosity is often one of the most important parameter when we describe the genetic data. Using this measure we explain the general trend in the structure of analysed populations – even is their history and future genetic structure is concerned. Low values of heterozygosity is influenced by small population size and processes of genetic drift e.g. bottlenecks effect. A lot of heterozygotes in population signify high genetic variability. When we compare the level of the observed and expected heterozygosity in balanced populations concerning random and open mating system (i.e. under Hardy-Weinberg equilibrium) and the observed heterozygosity level is higher than the expected one, we can presume the gene flow via alien pollen outside of population. If the observed heterozygosity is lower than expected one we can assume that some inbreeding processes may occur in the population.

The interpopulational variation is described very often as G_{ST} [9] used as equivalent of F_{ST} in G_{ST} statistics [10, 11], and it enables to assess the distance for each population from other populations (3).

$$G_{ST} = \frac{H_T - H_S}{H_T} \tag{3}$$

H_T-interpopulation heterozygosity

H_S-intrapopulation heterozygosity

The F_{ST} parameter called as fixation index is the measure of proportion of the total genetic variance within subpopulations in relation to the total genetic variance (4). The values of this parameter can range from 0 to 1. High F_{ST} implies a considerable degree of differentiation among populations.

F_{IS} (inbreeding coefficient) is the proportion of the variance in the subpopulation. High F_{IS} implies a considerable degree of inbreeding (5). Values can range from-1 (outbred) to+1 (inbred).

$$F_{ST} = (H_T - H_S) / H_T \tag{4}$$

$$F_{IS} = (H_S - H_I) / H_S \tag{5}$$

H_T-total heterozygosity for a population

H_S-heterozygosity within a subpopulation

H_I-heterozygosity of an individual

2.2.2. Software

One of the oldest programs enabling computing of DNA markers analysis data is BIOSYS-2. This program was elaborated to help biochemical population geneticists to describe the analysis of electrophoretically detectable allelic variation. It can be utilized to study allele frequencies and genetic variability measures, to test deviation of genotype frequencies from Hardy-Weinberg, expectations, to calculate F-statistics, to perform heterogeneity of chi-square analysis, to calculate a variety of similarity and distance coefficients, and finally to construct dendrograms using among others cluster analysis procedures. The program, documentation, and test data are available from the authors [12].

The statistical analysis of the alleles consists in calculation of the genetic parameters estimated by Nei [9, 13], i.e. expected heterozygosity (H_E), observed heterozygosity (H_O), observed number of alleles (A_O) and population differentiation parameters (H_S, H_T, F_{ST}). Those parameters may be calculated with the programs like GENEPOP software 3.2a [14], GenALEx [5] or

ARLEQUIN (http://lgb.unige.ch/arlequin/). The spatial correlations may be evaluated with SPAGeDi v.1.2 [15] and the genetic distances estimated according to Nei [13]. Another one interesting software enabling the analysis of DNA data markers is POPGENE [16]. The current version of POPGENE is designed specifically for the analysis of co-dominant and dominant markers using haploid and diploid data. The software performs most types of data analysis encountered in population genetics and related fields. It can be used to compute summary statistics, including: allele frequency: estimates of gene frequencies at each locus from raw data, effective number of alleles per locus, percentage of all polymorphic loci, observed and expected homozygosity, Shannon Index, gene diversity Nei's [9], F-Statistics, gene flow from the estimates of G_{ST} or F_{ST} and many others parameters.

All those programs represent good tool for population genetics analysis and simulations, including: Hardy-Weinberg Equilibrium (HWE), multiple allele and loci inheritance, natural selection, genetic drift, migration, mutation and inbreeding.

3. Results presentation Genetic variation characteristics of *Fagus sylvatica* L. as an example of utilization of capillary electrophoresis method on the basis of nuclear and chloroplast DNA markers

3.1. Object of the study

European Beech (*Fagus sylvatica* L.) is one of the most important forest tree species in Poland. Beech forests cover about 5.6 % of forest area [17]. The most typical beech forest tree populations are formed at the lower forest belt in Carpathians and Sudety Mountains on the South and at the moraine landscape of Pomeranian Lake District of the North of the country. In Poland beech attains its north-eastern limit of natural range [18, 19]. Varying environmental conditions have resulted in a great number of ecotypes and populations which are characterized by various ecological requirements [20, 21]. The growth of beech stands outside the natural beech limit indicates that species possess potentially much wider range [22, 23].

Present genetic structure of beech populations in Poland was formed by many different factors, not only environmental and genetic ones but also anthropogenic [24, 25, 26, 27]. Recent investigations of beech variation in Poland performed with isoenzyme study [28, 29, 30, 31], showed high genetic diversity, similar to other neighboring European populations, slight decrease of average number of alleles per locus and lower level of differentiation towards the North of the natural range limit, which generally confirm the migration paths after glaciations. The present paper describes the genetic structure within one generation, i.e. mother and progeny beech stands in Poland assessed with chloroplast and nuclear DNA markers.

There were investigated six beech populations representing natural beech range in Poland. The populations were classified according to phytosociological characteristics to the following plant associations: *Galio-odorati-Fagetum* (Gryfino and Kwidzyn), *Dentario glandulosae-Fagetum* (Bieszczadzki National Park), *Luzulo-luzuloides-Fagetum* (Suchedniów, Tomaszów),

Dentario enneaphyllidis-Fagetum (Zdroje) - Fig. 7. The genetic structure of these populations was analysed.

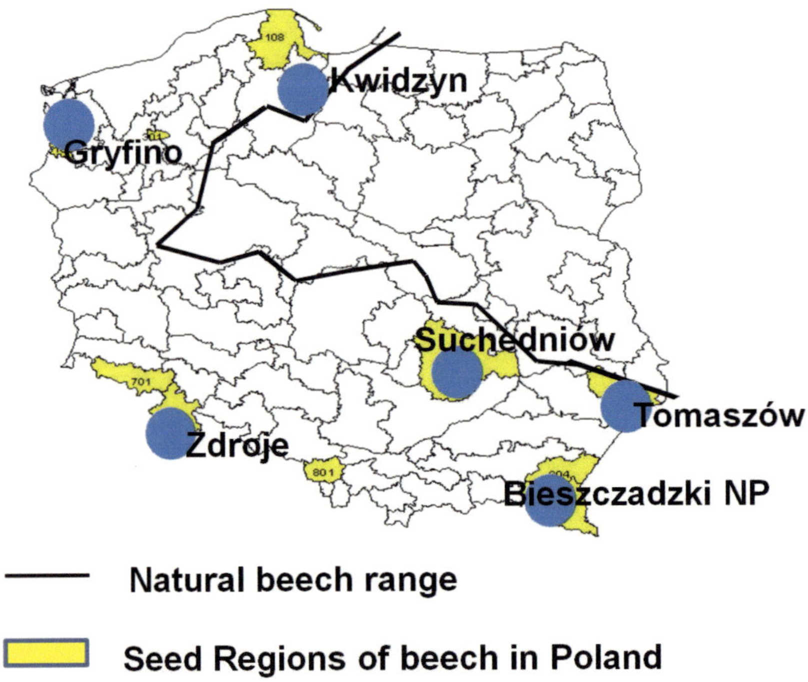

Figure 7. Localisation of investigated Polish European beech populations.

3.2. Methods

The genetic variation and differentiation of mother stands and their open-pollinated progeny were characterised on the basis of nuclear microsatellite markers, i.e. FS1-03, FS1-25, FCM5, mcf5, mcf11 [32, 33], as well as chloroplast DNA markers: ccmp4, ccmp7, ccmp10, according to reference [34, 35]. Thirty individuals per one generation (mother, progeny stands) in every provenance were investigated.

The extraction of total DNA from the leaves was performed using Qiagen DNeasy™ Plant Minikit according to the manufacturer instruction (Qiagen). The quality and purity of DNA were analyzed on 1% agarose gel electrophoresis and via absorption in 230, 260 and 280 nm in NanoDrop® spectrophotometer (Wilmington, USA). DNA samples were analyzed with DNA capillary electrophoresis in Beckman Coulter® sequencer, and analyzed using the software CEQ™8000 Genetic Analysis System v 9.0 (Fullerton, USA).

Parameters of genetic diversity (H_S and H_T) and differentiation (G_{ST}) were counted and compared between mother and progeny generation according to Nei [36, 11] in PopGene 1.32 software [16].

http://dx.doi.org/10.5772/59197

3.3. Results

3.3.1. Quality and quantity of the analyzed DNA

The very high quality and purity of DNA were assessed on the basis of the ratio of absorbance at 260 and 280 nm. A ratio about ~1.8 was typical for most of the samples (Fig. 2), and proved a high purity of the extracted DNA. A ratio ~2.0 is generally accepted as RNA-free. If the ratio is lower in either case, it indicates the presence of contaminants. The good quality of isolated DNA from samples was confirmed also by measurement of absorbance at 230 nm wavelength. The quantity of the genomic DNA samples balanced between 35 up to 160 ng.µl-1 and was fully appreciated to perform next steps of the DNA analysis procedure.

3.3.2. Genetic structure based on nuclear DNA markers

As far as nuclear microsatellite markers of Polish beech stands were concerned, different fragment-variants of haplotypes, i.e. from 73 to 348 base-pair size were distinguished. In investigated trees both heterozygous and homozygous alleles were found (Fig. 5).

The FS1-03 locus was the most polymorphic, i.e. exhibited 27 allele variants: 73 up to 154 bp DNA fragments, while the smallest number of variants was observed in case ofmcf-11locus – 11 alleles variants with DNA fragment length from 315 up to 348 bp. Usually the observed number of alleles per locus was higher for mother stands comparing to progeny stands except practically only population Tomaszów. For instance, mother trees presented mean observed number of alleles per locus 9.4, comparatively to 7.4 found in progeny from Tomaszów population (Fig. 8).

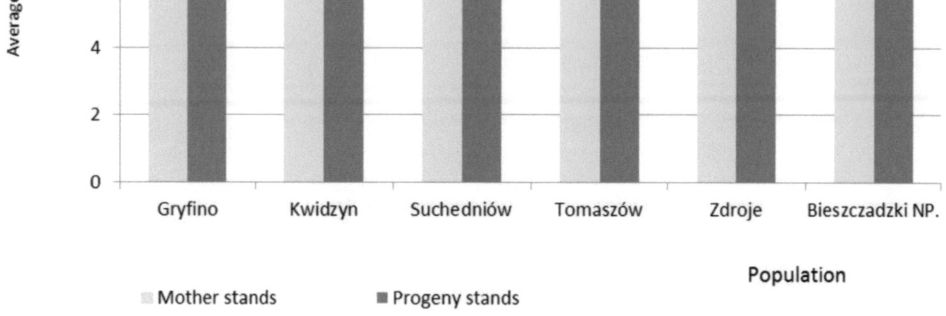

Figure 8. Average number of alleles in populations of investigated mother and progeny *Fagus sylvatica* L. stands.

Mean gene diversity among and within all studied mother stands (Tab. 1) were more or less at the same level (H_T=0.8291, H_S=0.7693, respectively) than mean genetic diversity in the progeny stands ($H_{T=}$0.8257, H_S=0.7672), as well as the gene diversity level of all mother and progeny stands were almost at the same level: $F_{ST=}$0.0587 and F_{ST}=0.0576 respectively (Tab. 1). The differentiation among studied Polish *Fagus sylvatica* population stands and their progeny can be explain that most genetic diversity resides within the stands.

Differentiation based on Nei [13] genetic distances was independent to geographical location of populations (Fig. 9).

Locus	**Mother Stands**			**Progeny Stands**		
	H_T	H_S	F_{ST}	H_T	H_S	F_{ST}
FS1-03	0.7165	0.6994	0.0404	0.7730	0.7284	0.0433
FS1-25	0.8855	0.8371	0.0489	0.8861	0.8354	0.0496
FCM-5	0.9072	0.8370	0.0590	0.8862	0.8275	0.0591
mcf-5	0.9127	0.8370	0.0683	0.8984	0.8309	0.0708
mcf-11	0.7236	0.6357	0.0766	0.6847	0.6140	0.0652
Mean	0.8291	0.7693	0.0587	0.8257	0.7672	0.0576
St. Deviation	0.1001	0.0955	-	0.0939	0.0966	-

Table 1. Genetic variation parameters of nuclear DNA markers microsatellite loci analyzed in *Fagus sylvatica* L. stands

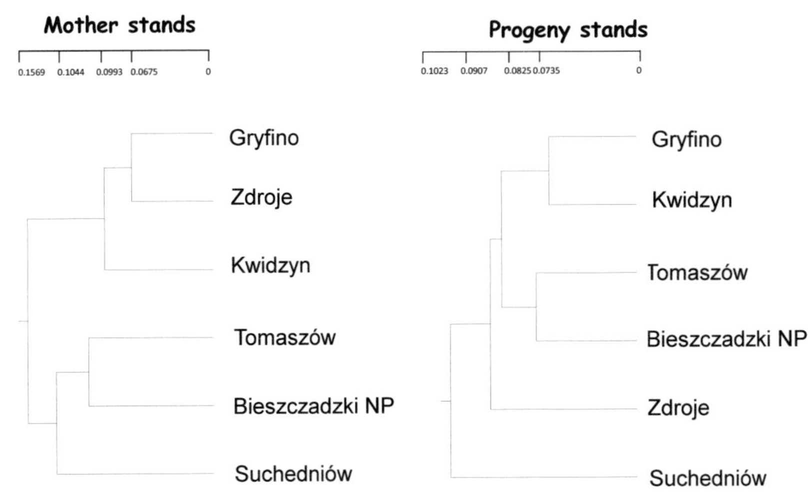

Figure 9. Genetic distances on the base of five SSR nuclear microsatellite markers

3.3.3. Genetic structure based on chloroplast DNA markers

As far as ccmp4 and ccmp10 microsatellite markers of Polish beech stands were concerned, different fragment-variants of haplotypes, i.e. from 116 to 152 base-pair size were distinguished. The ccmp7 locus was the most polymorphic, i.e. exhibited 8 allele variants: 144, 145, 147, 148, 149, 150, 151 and 152 bp. Nevertheless, some populations e.g. Bieszczadzki NP showed less polymorphism in the locus ccmp7, with only one 147 bp variant present, similarly to the previous study performed on Polish beech [38]. Another two loci, ccmp4 and ccmp10, shared the same range of 5 allele-size variants of 116, 117, 118, 119 and 120 bp. Generally, mother beech trees had more variable allele, than the progeny trees of the same provenance. For instance, mother trees presented alleles from 117 to 119 bp, comparatively to only one allele of 119 bp found in progeny from Bieszczadzki NP population. Mean gene diversity among and within all studied mother stands (Tab. 2) were slightly higher (H_T=0.4916, H_S=0.3606 respectively) than mean genetic diversity in the progeny stands (H_T=0.4600, H_S=0.3375). The gene diversity level of all mother and progeny stands were almost at the same level: G_{ST}=0.2666 and G_{ST}=0.2663 respectively (Tab. 2). The overall haplotypic differentiation among studied Polish *Fagus sylvatica* populations was quite low (G_{ST}=0.016), which means that most genetic diversity resides within the stands.

Differentiation based on Nei [13] genetic distances was independent to geographical location of populations (Fig. 10).

Locus	Mother Stands			Progeny Stands		
	H_T	H_S	G_{ST}	H_T	H_S	G_{ST}
ccmp4	0.5053	0.3058	0.3947	0.3957	0.3042	0.2313
ccmp7	0.3228	0.2667	0.1738	0.4650	0.2767	0.4050
ccmp10	0.6468	0.5092	0.2128	0.5193	0.4317	0.1688
Mean	0.4916	0.3606	0.2666	0.4600	0.3375	0.2663
St. Deviation	0.0264	0.0169	-	0.0038	0.0068	-

Table 2. Genetic variation parameters of chloroplast DNA markers microsatellite loci analyzed in *Fagus sylvatica* L. stands

3.4. Discussion

The DNA markers constitute powerful tool when gene variability of forest trees is assessed. The complicity of genome of these organisms makes impossible in most cases to obtain information about particular gene structure or variability. Nowadays, microsatellite markers analysed via CE technology overcome this difficulty, enabling the genetic variation study of the given organism, but only at the level of non-coding DNA regions.

The great advantage of the CE method is that all steps of allele fragment genotyping and DNA sequencing analysis are performed automatically in one integrated system. The laboratory tasks are then concentrated on high purity DNA molecules isolated from plant tissue and on appropriate PCR amplification procedures.

The data of CE application to study the SSR markers in European beech stands in Poland revealed: high genetic diversity of beech, similar like in other neighboring European popula-

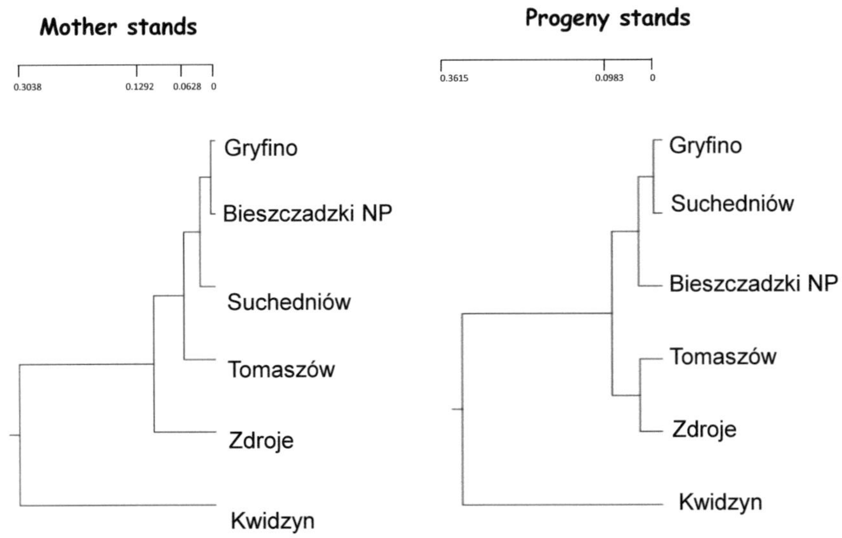

Figure 10. Genetic distances on the base of three chloroplast ccmp markers

tions, slight decrease of average number of alleles per locus and level of differentiation towards the North of the natural range limit, which generally confirm the migration paths after glaciations but it is not the basis to distinguish geographic regions. More cpDNA variation in the chloroplast ccmp4, ccmp7 and ccmp10 loci (G_{ST}=0.810) was reported for other 400 European beech populations [37]. Nevertheless, some populations e.g. Bieszczadzki NP showed less polymorphism in the locus ccmp7, with only one 147 bp variant present, similarly to the previous study performed on Polish beech [38].

In most cases, higher level of genetic variation was found within the investigated beech populations, predispose them to higher genetic tolerance against harmful environmental factors-reference [39].

4. Conclusions

In recent years the development and dissemination of new molecular methods based on CE method has remarkably increased. The use of automatic machines (sequencers) enables the investigation not only of DNA and RNA structure, but also gene bank creation for the forest tree species in order to achieve the appropriate forest stand management level. The CE gives very good resolution in separation of DNA molecules at the level of single nucleotide base-pair. Moreover it helps to overcome the problem of homoplasy in population genetics and offer cheap and fast results in sequencing of a small amount of samples. Concurrent development of software and computerisation offered the possibility of complete automation of the sample processing data, including the transfer of data and results.

Application of CE electrophoresis method is useful tool in forest trees genetic diversity assessment, despite of their limits. It is possible to use them on wild scale, especially in:

- Genetic characteristics of different forest reproductive material of natural and artificial stands e.g.
  - Mother stands and progeny stands gene flow
  - Seed orchards gene flow
  - Gene banks representation of populations assessment
- Solving of particular problems in case of:
  - Clonal/pedigree identification/selection process
  - Pollen contamination
  - Mating system
  - Patterns of gene flow
- Good tools to support conservation and management of forest genetic resources e.g. to support following activities:
  - Selection and protection of ecotypes
  - To asses initial gene pool for needs of effective gene conservation measures
  - To present selection processes of forest stands to maintain rich natural diversity
  - Prevention of illegal logging procedure thanks to the polymorphic SSR markers assessed with CE technique.

Acknowledgements

Special thanks are addressed to Jolanta Bieniek and Malgorzata Borys for their laboratory assistance in CE technique.

Author details

Justyna A. Nowakowska and Malgorzata K. Sulkowska*

*Address all correspondence to: m.sulkowska@ibles.waw.pl

Forest Research Institute, Sekocin Stary, Braci Lesnej, Raszyn, Poland

References

[1] Camilleri P. Capillary Electrophoresis: Theory and Practice. R. Weinberger Ed., CRC Press; 1997.

[2] Hawkins T.L., McKernan K.J., Jacotot L.B., MacKenzie B., Richardson P.M., Lander E.S. DNA sequencing – a magnetic attraction to high-throughput genomics. Science 1997; 276: 1887-1889.

[3] Fakruddin M., Mannan K.S., Chowdhury A., Mazumdar R.M., Hossain M.N., Islam S., Chowdhury M.A. Nucleic acid amplification: Alternative methods of polymerase chain reaction. Journal of Pharmacy and Bioallied Sciences 2013; 5: 245-252.

[4] Dolník V. DNA sequencing by capillary electrophoresis (review). Journal of Biochemical and Biophysical Methods 1999; 41: 103-119.

[5] Peakall, R., Smouse P. E. GenAlEx 6.5: genetic analysis in Excel. Population genetic software for teaching and research–an update. Bioinformatics 2012; 28: 2537–2539.

[6] Sambrook J., Russell D.W. Molecular cloning. A laboratory manual. Cold Spring Harbor Laboratory Press, Cold Spring Harbor, New York, USA. 2001; Vol. 1 and 2.

[7] Sanger F., Nicklen S., Coulson A.R. DNA sequencing with chain-terminating inhibitors. Proceedings of the National Academy of Sciences USA 1977; 74(12): 5463-5467.

[8] Crow, J.F., Kimura, M. Introduction to Population Genetics Theory. Harper and Row, New-York; 1970.

[9] Nei, M. Analysis of gene diversity in subdivided populations. Proceedings of the National Academy of Sciences, USA. 1973; 70(12, Pt. 1), 3321-3323.

[10] Wright, S. Evolution and the Genetics of Populations, Vol. II. The Theory of Gene Frequencies. University of Chicago Press, Chicago; 1969.

[11] Wright, S. Evolution and the Genetics of Populations, Vol. IV. Variability Within and Among Natural Populations. University of Chicago Press, Chicago; 1978.

[12] Swofford, D.L., Selander, R.B. (1997). Biosys–2. User Manual. University of Illinois; 1997.Available from http://fortinbras.us/biosys2/ (accessed 5 February 2009).

[13] Nei, M. Estimation of average heterozygosity and genetic distance from a small number of individuals. *Genetics* 1978; 89: 583-590.

[14] Raymond, M., Rousset F. GENEPOP (version 1.2): population genetics software for exact tests and ecumenism. Journal of Heredity 1995; 86: 248-249.

[15] Hardy O.J., Vekemans X. Spagedi: a versatile computer program to analyse spatial-genetic structure at the individual or population levels. Molecular Ecology Notes 2003; 4: 618-620.

[16] Yeh, F.C., Boyle, T.B.J. Population genetic analysis of co-dominant and dominant markers and quantitative traits. Belgian Journal of Botany 1997; 129: 157.

[17] Forests in Poland 2013. Warszawa, Centrum informacyjne Lasów Państwowych 2013;http://www.lasy.gov.pl/publikacje/in-english/forests-in-poland-2013/view (access 11 March 2014).

[18] Szafer W., Pawłowski W. Szata roślinna Polski. [The look–description of vegetation.] Warszawa, PWN; 1972.

[19] Boratyńska K., Boratyński A. Systematyka i geograficzne rozmieszczenie. [Systematic and geographic distribution.] In: Białobok S. (ed.): Buk zwyczajny *Fagus sylvatica*. [European beech *Fagus sylvatica*] Warszawa – Poznań: PWN; 1990. p 27-73

[20] Dzwonko, Z. Ekologia. [Ecology] In: Białobok S. (ed.): Buk zwyczajny *Fagus sylvatica*. [European beech *Fagus sylvatica*] Warszawa – Poznań: PWN; 1990. p237-328.

[21] Giertych M. Genetyka. [Genetics] In: Białobok S. (ed.): Buk zwyczajny *Fagus sylvatica*. [European beech *Fagus sylvatica*] Warszawa – Poznań: PWN; 1990. p193-237.

[22] Brzeziecki B. Skale nominalne wymagań klimatycznych gatunków leśnych. [Nominal climatic requirements forest species range.] Sylwan 1995; 139(3): 53-65.

[23] Tarasiuk S. Buk zwyczajny (*Fagus sylvatica* L.) na obrzeżach zasięgu w Polsce. Warunki wzrostu i problemy hodowlane. [European Beech (*Fagus sylvatica* L.) in its marginal sites in Poland. Growth conditions and silvicultural problems.] Warszawa, Fundacja „Rozwój SGGW"; 1999.

[24] Szafer, W. The significance of isopollen lines for the investigation of the geografical distribution of trees in the Post-Glacial period. Bulletin de l'Academie Polonaise des Sciences et des Lettres, Serie B 1935; 1: 235-239.

[25] Huntley, B., Birks, H.J.B. An atlas of past and present pollen maps for Europe: 0–13000 years ago. Cambridge Univ. Press; 1983

[26] Ralska-Jasiewiczowa, M. Isopollen maps for Poland: 0–11,000 years B.P. New Phytology 1983; 94: 133-175.

[27] Hazler, K., Comps, B., Šugar, I., Melovski, L., Tashev, A., Gračan, J. Genetic structure of *Fagus sylvatica* L. populations in Southeastern Europe. *Silvae Genetica* 1997; 46(4): 229-236.

[28] Sułkowska M. Isoenzyme analysis chosen provenances of beech (*Fagus sylvatica* L.) on the provenance trial in Bystrzyca Kłodzka. Sylwan 2002; 146 (2): 129-137.

[29] Gömöry, D., Paule, L., Schvadchak, M., Popescu, F., Sułkowska, M., Hynek, V., Longauer, R. Spatial patterns of the genetic differentiation in European beech (*Fagus sylvatica* L.) at allozyme loci in the Carpathians and adjacent regions. *Silvae Genetica* 2003; 52(2): 78-83.

[30] Sulkowska, M. Genetic and ecotype characterization of European beech (*Fagus sylvatica* L.) in Poland. *Acta Silvatica et Lignaria Hungarica* 2010; 6: 115-122.

[31] Sułkowska, M., Gömöry, D., Paule, L. Genetic diversity of European beech in Poland estimated on the basis of isoenzyme analysis. *Folia Forestalia Polonica* – series A Forestry 2012; 54(1): 48-55.

[32] Pastorelli R., Smulders M. J. M., Westende Van't W. P. C., Vosman B., Giannini R., Vettori C. Vendramin G. G. Characterization of microsatellite markers in *Fagus sylvatica* L., *Fagus orientalis* Lipsky. Molecular Ecology Notes 2003; 3, 76-78.

[33] Vornam B., Decarli N., Gailing O. Spatial distribution of genetic variation in natural beech stands (*Fagus sylvatica* L.) based on microsatellite markers. Conservation Genetics 2004; 5: 561-570.

[34] Grivet, D., Heinze, B., Vendramin, G. G., and Petit, R. J. Genome walking with consensus primers: application to the large single copy region of chloroplast DNA. Molecular Ecology Notes 2001; 1(4), 345-349.

[35] Weising K, Gardner RC. A set of conserved PCR primers for the analysis of simple sequence repeat polymorphisms in chloroplast genomes of dicotyledonous angiosperms. Genome 1999; 42(1):9-19

[36] Nei, M. Genetic distance between populations. American Nature 1972; 106: 283-292.

[37] Magri, D., Vendramin, G. G., Comps, B., Dupanloup, I., Geburek, T., Gömöry, D., Latałowa, M., Litt, T., Paule, L., Roure, J. M., Tantau, I., van der Knaap, W. O., Petit, R., Beaulieu, J. L. A new scenario for the Quaternary history of European beech populations: paleobotanical evidence and genetic consequences. New Phytology 2006; 171(1): 199-221.

[38] Nowakowska J.A., Oszako T. Health condition and genetic differentiation level of beech in the Siewierz Forest District assessed with cpDNA markers. Sylwan 2008; 152 (9): 11-20.

[39] Starke, R., Ziehe, M., Müller-Starck, G. Viability selection in juvenile populations of European beech (*Fagus sylvatica* L.). Forest Genetics 1996; 3(4): 217-255.

Section 3

Gel Electrophoresis

Chapter 5

The Use of Electrophoresis for the Study of Saliva Involvement in Ingestive Behavior

Elsa Lamy, Martine Morzel, Lénia Rodrigues, Cristina Pinheiro, Ana Rodrigues Costa, Célia Miguel Antunes, Orlando Lopes and Fernando Capela e Silva

Additional information is available at the end of the chapter

http://dx.doi.org/10.5772/59290

1. Introduction

Whole saliva contains a wide variety of proteins and peptides, of glandular or blood origin, with diverse functions, namely proteins that participate in teeth and oral tissues protection, as well as proteins related to ingestive and digestive processes [1]. Several studies, in animals and humans, present evidences that saliva is involved in eating behaviour [e.g. [2–5]]. The link between saliva composition and oral perception is increasingly reported. Several studies refer that oral sensations, such as astringency, result from an interaction between food constituents (e.g. polyphenols) and salivary proteins [6]. Moreover, the involvement of saliva in taste sensitivity has been also considered and an example of this relationship is the correlation established between taste dysfunction and reduced levels of the salivary protein carbonic anhydrase VI [7]. Recently, some studies added evidences that protein saliva composition is involved in taste by associating salivary proteins to fat perception and liking [8], as well as to sensitivity for the bitter tastes of 6-n-propylthiouracil (PROP) [9] and caffeine [10]. Besides the involvement of saliva in oral perception, which can greatly influence food choices, the effect of this fluid in ingestive behaviour can also be considered by the role of certain salivary proteins in digestive processes and/or in the regulation of energy intake: one example is salivary α-amylase, which is involved in the digestion of starch in the mouth; another example is the presence in saliva of proteins involved in appetite/satiety regulation such as leptin [11] and ghrelin [12].

Electrophoresis has been applied for salivary protein separation for years and progresses have been made until now. Several studies report salivary protein separation according to their mass, isoelectric point and both parameters (two-dimensional electrophoresis – 2DE), both for human [e.g. [10]] and animal saliva [e.g. [13]]. The application of electrophoretic techniques to saliva samples, together with advances in mass spectrometry, for protein identification, resulted in an increased interest in this biological fluid as a source of biomarkers. In fact, saliva collection has the great advantage of being relatively easy, cheap and non-invasive to perform, presenting proteins also existent in other body fluids such as blood. Besides Sodium Dodecyl Sulfate Polyacrylamide Gel Electrophoresis (SDS PAGE) and 2DE, salivary protein and peptide separation have been described to be achieved through capillary electrophoresis (CE). The possibility of work with small sample volumes and to perform high-resolution analysis is a great advantage of CE [14]. Although only few studies about salivary protein/peptide profiles related to ingestive behaviour have used CE, we believe that this area of research can gain from the use of this approach.

Electrophoresis has also the potential of being used in paraffin embedded tissue samples [15]. This possibility has the great benefit of allowing access to already acquired tissue samples which can be readily correlated with histological parameters. Moreover, it provides access to tissue that would be either difficult to collect prospectively in a timely manner or unlikely to be available as fresh samples. For salivary composition analysis it may have the potential of allowing the study of individual salivary glands composition, in situations where no individualized glandular saliva collection was possible. As such, this may have the potential of aiding in the comprehension of salivary gland regulation and differential contribution for total salivary composition.

This chapter gives an overview of the use of electrophoresis in studies about the involvement of saliva in ingestive behaviour. Studies about the role of salivary protein composition in oral sensations and taste sensitivity will be reviewed. The particularities in the use of electrophoresis in saliva samples will be discussed, as well as the advantages and limitations of the technique for the analysis of this body fluid. The major limitations in the electrophoretic study of salivary proteins, such as the presence of high abundant proteins (impairing the study of scarce ones), the high content of mucins, the high proteolytic activity, among others, will be critically commented based also on our experience. Moreover, the promising use of capillary electrophoresis, as well as electrophoresis of paraffin-embedded tissue samples, for the study of salivary secretion will be proposed. Finally, non-electrophoretic techniques that can give complementary information will be presented, accentuating the importance of integrated approaches in the study of saliva and salivary secretion.

2. The involvement of salivary proteins in food perception and choices

2.1. Taste sensitivity

Taste perception occurs when a sapid molecule (or tastant) activates a taste receptor of the tongue. Taste receptors are located mainly on cells of specialised structures, the taste buds, which are themselves situated in the clefts of fungiform, circumvallate or foliate papillae. Five basic tastes are commonly admitted: sweetness, saltiness, bitterness, sourness and umami. In addition, discovery of receptors to free fatty acids on the human tongue [16,17] has prompted suggestions that fat could be a sixth taste.

The tongue is constantly lined and hydrated by the oral fluid made of secreted saliva, crevicular fluid, microorganisms etc. Therefore, saliva is intuitively associated to the gustatory function. This assumption applies even more prominently to saliva secreted by the von Ebner's glands (VEG). These minor salivary glands secrete their saliva at the bottom of the clefts of circumvallate and foliate papillae, i.e. in near vicinity to the taste buds. In other words, taste buds bathe in saliva from the VEG, a fact that attracted interest in this type of saliva as playing a role in taste perception. To our knowledge, systemic proteome analysis of VEG saliva has never been performed. However, by studying VEG from rats [18,19] and human subjects [20], two specific proteins have been identified. These are the VEG protein, which is also known as lipocalin 1, and Ebnerin. Lipocalin 1 was found in VEG saliva of human subjects [21]. This protein presents some sequence homology with transporters of hydrophobic molecules, and therefore some authors have proposed that it may help in concentrating and delivering hydrophobic tastants to the specialised gustatory cells [18]. However, in a sensory-based study, lipocalin 1, although measured, was not pointed as an important salivary factor correlated with detection threshold to the taste of oleic acid [22]. Bläker et al. (1993) [20] suggested another function for lipocalin 1, namely that it has a protective effect of taste structures against the detergent action of fatty acids. Concerning Ebnerin, it has been found only in rats, more specifically released into the clefts of circumvallate papillae [19]. The putative function attributed to this protein is to carry growth factors which would be expressed in VEG [19]. To conclude with VEG saliva, despite the proximity of VEG and taste buds, relatively little evidence is provided for involvement of VEG saliva proteins in taste perception.

Salivary proteins or peptides have been related to taste perception, acting at different levels (Figure 1). Some of these proteins/peptides found in whole saliva have been suggested to act as factors implicated in the growth and renewal of taste buds. For example, carbonic anhydrase 6 (CA6 or gustin) levels are reduced in hypogeusic patients [23] and such patients also present taste bud anatomical abnormalities [24]. CA6 was therefore considered as a trophic factor involved in taste bud growth and development. Leinonen et al. [25] later suggested that CA6 may also have an anti-apoptotic action on taste buds by locally regulating pH. More recently, the perception of the bitter compound 6-n-propylthiouracil (PROP) was also associated to the protein CA6. However, it was not the quantitative expression of the protein that was different between PROP tasters or nontasters, but rather the ability of CA6 to bind zinc which conditions its functionality. This ability differed according to genetic polymorphism [26], although some controversy appears to exist, with a recent study lacking to obtain similar results [27]. Another protein, metalloproteinase MMP-3, is under-expressed in saliva from subjects with taste disorders and may serve in continuous regeneration of taste buds [28]. Finally, the epidermal growth factor (EGF) secreted by the parotid and submandibular glands might assist in maintaining the morphological integrity of taste buds. In rats, for instance, removal of major salivary glands alters greatly the appearance of the taste buds of fungiform papillae but oral administration of EGF restores a normal morphology [29]. Whether salivary EGF is linked to taste in humans has not been demonstrated to our knowledge.

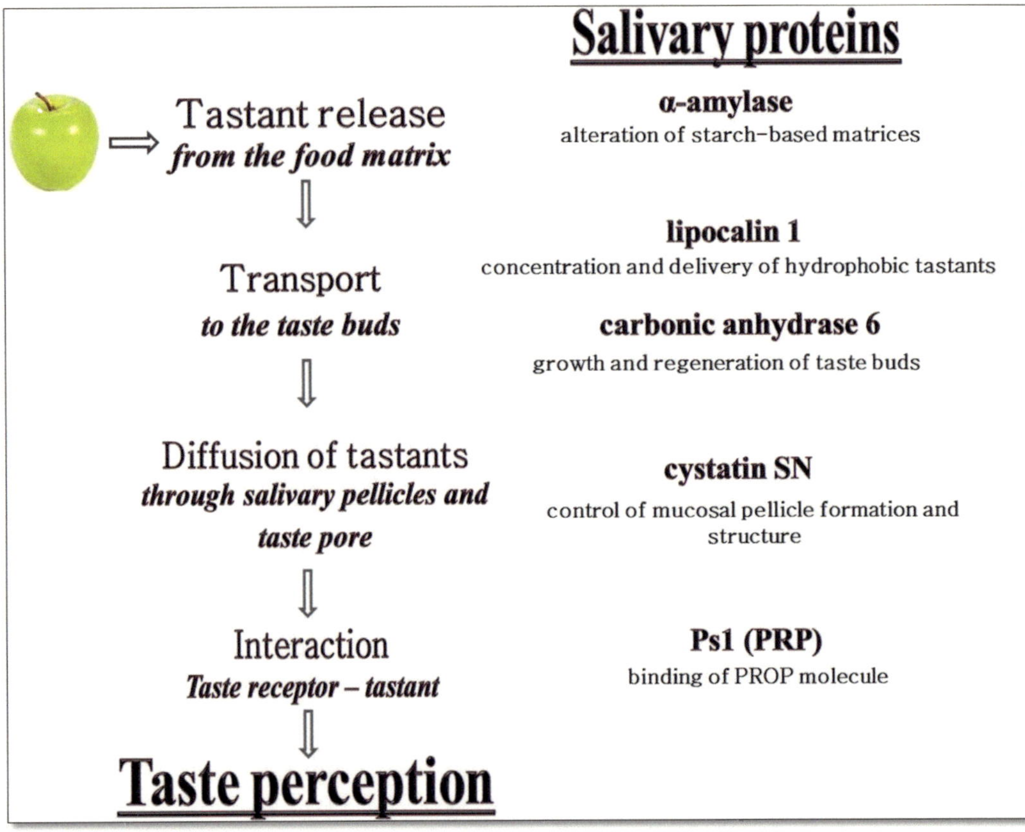

Figure 1. From food to taste perception: examples of salivary proteins involved in taste function (right – examples of salivary proteins potentially involved in each of the phases of taste perception described on the left)

Another impact of salivary proteins on taste perception resides in their direct physico-chemical interaction with taste molecules, which modifies the availability of tastants to the taste receptors. For example, Wada et al. (2010) [30] have shown that histatin 5 concentration was significantly lower in subjects hypersensitive to the taste of quinine, and the authors further demonstrated that histatin 5 could bind quinine. For a given quantity of quinine, subjects with lower levels of histatin 5 had therefore more free quinine which could interact with their taste receptors, thereby increasing their sensitivity. Another example of salivary protein – tastant interaction has been suggested by [31] based on the demonstration by ^{1}H-NMR that the PROP molecule could interact with the arginine and lysine amino-acids. This investigation followed a study where top-down analysis of saliva of supertasters *vs* nontasters of PROP revealed that saliva of supertasters had higher levels of basic proline-rich proteins (PRPs), in particular the Ps1 protein which contains a high proportion of arginine and lysine [9]. In this particular case, higher proportion of Ps 1 and the consequent binding of PROP would confer to the molecule a new conformation, which would facilitate its access to its taste receptor. Finally, the interaction of oleic acid with the salivary protein zinc-α-2glycoprotein (ZAG) has been suggested, on the basis that this protein was over-expressed in a group of subjects hypersensitive to the taste of this fatty acid [32]. This protein may, for example, act as a solubilizer of oleic acid in saliva.

In addition to these two mechnisms (growth or protection of taste buds/physico-chemical interactions with tastants), it is also possible that the structured biological layer covering the tongue surface, and made mainly of salivary proteins, may contribute to modulate accessibility of tastants to the taste receptors. This hypothesis has been formulated by Dsamou et al. (2012) [10] and deserves further investigations. In this context, one should focus in particular on the "taste pore material" which is the biological layer covering directly the taste buds, described for example by Matsuo (2000) [33].

2.2. Mechanical sensations (astringency)

Physical properties of foods are also important factors in consumer appreciation and enjoyment of food products. Salivary proteins are known to modulate food tactile characteristics by interacting with food constituents. The involvement of salivary proteins in food perception started to be studied in terms of their effect in astringency development. Astringency is described as "the complex of sensations due to shrinking, drawing or puckering of the epithelium as a result of exposure to substances such as alums or tannins" by the American Society for Testing of Materials [34]. Astringent molecules are mainly plant-based products, namely polyphenols, among which tannins [35], but also acids and metal alums [6] and other dehydrating agents such as alcohols [36], which can also create this subjective oral feeling.

Astringency has been considered as an aversive quality responsible for the lowering acceptability of some plant food products [37,38]. Two possible mechanisms were proposed to explain this oral sensation. The oldest belief is that astringency is due to the precipitation of salivary proteins by astringent substances, with such precipitates increasing friction between mouth surfaces and stimulating mechanoreceptors [39]. On the other hand, a number of studies suggest that astringency is the result of modifications in the viscous elastic properties of glycosylated proteins, which result in the rupture of the lubricating saliva film that lines the oral cavity [40–43]. Nayak and Carpenter [43] proposed a more integrative view, suggesting a two-step interaction between salivary proteins and polyphenols (and/or other astringent substances). According to this model, saliva is considered to be composed of two different phases, a thin dynamic film coating the internal oral surfaces and an adsorbed layer of proteins on the hard and soft tissues. In the first step of interaction, the astringent molecules may bind the proteins from the dynamic film with the highest binding affinity. In the second step, the remaining astringent molecules, not bound in the first step, can interact with the adsorbed glycoprotein layer, with the consequent oral cavity loss of lubrication and astringency development.

Different families of salivary proteins have been considered as having a specific role in astringency of polyphenols, due to its high affinity for these compounds: PRPs [44], histatins [45], statherins and cystatins [46]. These salivary proteins represent a considerable part of the saliva total protein content and have important biological functions, in this oral fluid, associated with calcium binding to enamel, maintenance of ionic calcium concentration (PRPs and statherin), antimicrobial action (histatins and cystatins), or protection of oral tissues against degradation by proteolytic activity (cystatins) [47]. The nature of the interaction between these salivary proteins and polyphenols depends on several factors, among which protein characteristics and the type of polyphenol. Salivary proteins such as acidic PRPs and statherins present lower selectivity towards polyphenol structures, comparatively to histatins and cystatins [46]. Among these, salivary PRPs were by far the most studied, being generally considered as the main family of salivary proteins involved in astringency. PRPs have an extended structure allowing them to have a high affinity to bind tannins [48]. Basic PRPs have been suggested as having a primary role in the prevention against the negative antinutritive and/or toxic effects of these polyphenols [37,49]. The induction in the secretion of basic PRPs by the regular consumption of tannins reinforced this hypothesis [50]. Additionally, glycosylated PRPs, which are thought to contribute to salivary lubrication [51] also interact with tannins [52,53] and, consequently, these are proteins with potential impact in astringency.

Mucins also seem to have a role in astringency, although some controversy exists among studies. These proteins are the main responsible molecules for the viscoelastic properties of saliva. Some authors reported the binding of mucins to polyphenols, and consequently an effect on astringency development [54]. On the other hand, other authors [e.g. [36]], observed that these proteins are precipitated by alum and acid, but not by polyphenols, suggesting different involvement according to the type of astringent molecule.

From all these different studies, with different suggestions about astringency mechanism, what appears to be unequivocal is the complexity of this oral sensation, with multiple mechanisms potentially involved in its development, and the participation of salivary proteins.

3. Usefulness of electrophoresis for ingestive behaviour analysis using saliva

3.1. Methodological issues related to the use of electrophoresis in saliva study

Accurate examination of salivary components requires optimal collection, processing and storage conditions and, as such, there is a need for standardized protocols [55,56]. Salivary secretion is mainly regulated by the two branches of the autonomic nervous system (both sympathetic and parasympathetic), and as such it presents some plasticity, changing in amount and composition according to the nature of the stimulus. Factors such as circadian rhythm [57], gender [58], drugs [59], exercise [60], eating [61], among others, result in variations in saliva composition. Based on that, it is important to define the protocol of collection method. Firstly, it must be determined if the collection of saliva will be performed without or with stimulation. According to the first, the most used procedure consists in allowing saliva to drip off the lower lip into a tube maintained on ice. Concerning stimulated saliva, it is frequently obtained after parafilm mastication, or after sour taste stimulation [62]. Advantages and disadvantages of each approach have been mentioned previously [63]. Another issue to be considered is the origin of collected secretions, i.e., whether it is glandular or whole saliva, since considerable differences exist in composition. Reported storage conditions of saliva samples have varied widely among studies, with no current consensus on optimal procedures. Saliva contains many salivary proteins that are processed by post-translational modifications (PTMs), namely glycosylation, phosphorylation, sulfation and proteolysis, these modifications being responsible for many of this fluid functions. Consequently, de-glycosylation, de-phosphorylation and proteolysis should be minimized, and different research groups have employed different methods with this purpose. For example, the addition of ethanol apparently allows the storage of saliva samples at room temperature (for a period of about two weeks) without considerable changes [64]. The use of protease inhibitors to prevent proteolysis was also referred as allowing saliva storage at 4°C, during approximately 2 weeks, without significant degradation [64]. Nonetheless, it is known that not even an inhibitor cocktail can prevent all protein degradation [65]. It is important to note that the ideal handling and storage procedures will greatly depend on the peptides/proteins of interest. Whereas working on ice for no longer than one hour, with subsequent storage of samples at -80°C, has been considered a safe and practical handling protocol [65,66], recovery of salivary immunoglobulin A (sIgA) has been suggested to be greater for short storage times at room temperature than when samples are stored at low temperatures [67]. Freeze-thaw cycles are also not desirable, inducing protein precipitation, in particular from low molecular mass components, resulting in loss of some proteins expression and particularly enzymatic activity [65]. In any case, little research has been directed on ways of minimizing degradative processes, and this is clearly needed [56].

Another methodological issue, in electrophoresis of salivary proteins, concerns the presence of particular proteins in high levels, obscuring the low abundant ones, for which analysis may be of interest. This is even more pertinent considering studies aimed at detecting physiological biomarkers, since most of them are present in saliva at low amounts. The protein α-amylase contributes to almost 60% of total salivary proteome [68] and, many times, its depletion is necessary. Salivary α-amylase depletion can be achieved through elution of samples from starch columns to reduce this protein amounts specifically [64,69].

Saliva contains a diverse array of proteins and particularities of each type should be considered when choosing the staining procedure. PRPs, which exist in considerable amounts in parotid saliva, contain reduced amounts of amino acids containing sulphur. Consequently, these proteins are not easily stained with silver [70]. On the other hand, these proteins stain violet-pink with Coomassie Brilliant Blue R-250, particularly when a destain protocol consisting in 10% acetic acid, instead of 10% acetic acid/10% methanolis used [71]. Phosphoserine, phosphothreonine and phosphotyrosine containing proteins are detected at low levels by using Pro-Q-Diamond [72]. Periodic-acid Schiff (PAS) is used for glycoprotein staining. Nevertheless this procedure presents the limitation of needing high levels of protein load [73]. When the amount of sample is limited and higher sensitivity is needed, Pro-Q-Emerald can be a good option for glycoprotein detection, since it can be approximately 50-fold more sensitive than PAS [74].

Some particular characteristics of saliva should also be taken into consideration for Western blotting. It is important to be sure that samples are being compared for an equal amount of total protein. Since by using this technique only the protein that react with the antibody is visualized, the existence of a control of the quantity of protein loaded is important. In some types of samples there are proteins which levels are proportional to the amount of protein loaded, and as such, they can function as internal controls. In these cases, the simultaneous use of primary antibodies for these proteins and the protein of interest may allow adjustments [e.g. 75]. Nevertheless, in saliva a protein which relative amount to total protein remains constant is not known. One way to circumvent this limitation is through the stainning of the membrane with a reversible stain (e.g. Ponceau), before incubation with the primary antibody, in order to visualize the several bands present in each lane [76, 77].

Although needing studies to elucidate that, it is possible that some salivary mucins and/or other proteins can irreversibly adsorb to cotton roll, and, consequently, being lost. Another possible drawback, when working with animal saliva is the low amount produced by some species. For example, for small rodents, which are frequently used as human and animal models, saliva collection is not easy without stimulation. The use of parasympathetic agonists, such as pilocarpine, is often used, since it induces the increase in the volume secreted without changing the relative amount of each protein. On the other way, the use of sympathetic agonists (e.g. pilocarpine) is also frequent, but in this case, when the amount of protein concentration is wanted. This type of stimulation induces the synthesis and secretion of proteins from salivary glands, but a relatively low volume [79].

Most of the gel-based approaches need further protein identification, most of the times by mass spectrometry. Although advances have emerged, in the last years, for some animal species, salivary proteins identification continues to be challenging, due to the lack of complete and annotated genome and protein sequences [81]. When this happens, the search in other related species databases is needed, increasing the possibility of to rise the number of false positive results.

3.2. The use of SDS PAGE and 2-DE in salivary proteome study related to ingestive behaviour

For both humans and animals, it has been observed that 2DE is the most popular technique for the global analysis and initial profiling of saliva, being used as a first step for protein separation, followed by mass spectrometry (MS) or tandem MS (MS/MS) [82]. 2DE simultaneously separates proteins according to their isoelectric points (*pI*) and molecular masses, enabling the visualisation and identification of several thousand proteins on a single gel. It is particularly useful for screening and comparing complex mixtures such as saliva samples. By opposition to the study of several different diseases, the use of saliva for biomarker discover in the field of nutrition only more recently started to emerge. One of the greatest advantages of 2DE in salivary protein study relates to its capacity of separating proteins with different PTMs, allowing their separate quantification. Many salivary proteins present different isoforms [e.g. [83]], with only some of them potentially related to taste sensitivity, making this individual isoform comparison essential.

Separation of proteins from mixed saliva, by SDS PAGE and 2DE, coupled to MS for protein identification, allowed a better understanding about the relation between saliva and taste. In 2DE profiles, the expression of some protein spots were observed to be differently changed following taste stimulation, these differences being dependent on the type of taste stimuli [84]. This suggests different adaptations of saliva to the five basic tastes. Concerning bitter taste, the use of these techniques also added a great contribution to advances of knowledge in this field. Until recently, only salivary CA6 was reported to influence taste sensitivity, whereas now it is known that other salivary proteins may also be involved. For example, by using 2DE based approaches, cystatin SN was identified as being differently expressed among groups of sensitivity to the bitter taste of caffeine [10]. In addition, recently, using SDS PAGE for separation of proteins from mixed saliva of 3-month-old-infants, differences in salivary profiles related to bitterness acceptance were reported: higher abundance of zinc-alpha-2-glycoprotein and CA6 was associated to a lower acceptance of this taste, whereas higher abundance of lactoperoxidase, S-type cystatins and prolactin inducible protein was associated to a higher acceptance [32,85].

Electrophoresis also proved to be useful in studies aimed at understanding astringency, namely in the search for the salivary proteins involved in the development of this oral sensation, as well as for assessing the astringency of some food compounds. Several different studies used SDS PAGE and 2DE to compare salivary protein profiles before and after tannin ingestion, and to identify salivary proteins interacting with these plant secondary metabolites [e.g. [86,87]]. Dinella et al. [88,89], by separating proteins through SDS-PAGE, presented evidences that differences among individuals in sensitivity for astringent stimuli relates to the differences they present in salivary protein composition after repeated stimulation, i.e., individuals in which tannin stimulation results in higher changes in salivary protein profiles are the ones who strongly detect astringency. The evaluation of polyphenol astringency was also demonstrated to be possible by using SDS PAGE to separate the salivary proteins present in human saliva after mixture with these plant secondary metabolites [54].

3.3. New perspectives for electrophoresis in saliva study

3.3.1. Capillary electrophoresis: potential and use in saliva study

Capillary electrophoresis (CE) is a separation method, performed in capillaries, in which analytes migrate through electrolyte solutions under the influence of an electric field. The electrophoretic velocity of the analyte will depend upon the magnitude of the electric field and its electrophoretic mobility, and upon the rate of electroendosmotic flow.

In the last years, some studies surged highlighting the potential of CE in several different areas of research [e.g. [14]]. A great advantage of this over the conventional electrophoretic techniques is that, although also based on the movement of molecules in an electric field, CE is not restricted to the separation of large molecules based on size or charge. As such, it allows also the separation of molecules that have low molecular mass and/or neutral charge. Moreover,

CE has the advantage of working with considerable small volumes of sample (from picoliters to nanoliters) [14], what in saliva study may be important.

The commonly separation modes by CE are capillary zone electrophoresis, capillary gel electrophoresis, capillary isoelectric focusing, micelar electrokinetic chromatography and capillary isotachophoresis. Capillary zone electrophoresis, also known as free solution capillary electrophoresis, is the simplest form of CE. The application of the sample occurs in a narrow zone or band, with the separation buffer surrounding it. With the application of the electric field, the migration of each component in the sample occurs according to its own mobility. In this method, non-charged molecules cannot be separated, since this will move at the velocity of electroosmotic flow.

Capillary gel electrophoresis is based on the use of a polymeric gel medium, inside the capillary, which acts as a molecular sieve in which smaller molecules migrate faster than larger ones. This method has the advantage of separating molecules that have similar charge-to-mass ratios, which in the absence of the gel medium (i.e. in free solution) would have similar electrophoretic migration rates. Capillary isoelectric focusing is a type of CE that offers significant advantages over isoelectric electrophoresis slab gels in terms of automation, separation speed and quantitation. Although at the beginning capillary isoelectric focusing was not readily accepted, due to the difficulties in isolating the focused zones for the detector, actually this separation technique presents the highest resolution of all charge based separation techniques, due to nowadays systems that allow the monitorization of the whole column at once [90]. Micellar electrokinetic capillary chromatography is an electrophoretic technique developed in the early 90's that extends the applicability to non-charged molecules, which cannot be separated using simple free solution CE [91]. This technique is based on the combined effects of: 1) the differential portioning of molecules between the aqueous buffer and the micellar phase; 2) the differential migration of ionic species. It consists in the separation as the result of the combined effect of the differential partitioning of molecules between the aqueous buffer and the micellar phase, as well as any differential migration of ionic species. The sample is inserted between a leading electrolyte, which as a higher mobility than ions in sample zone, and a terminating electrolyte, with lower mobility, relatively to these last. The separation achieved through capillary isotachophoresisis based on differences in the velocities of analyte ions within the sample zone.

CE can help to circumvent some limitations in the study of saliva. One constraint of searching for salivary protein biomarkers is that they may be massively diluted, requiring highly sensitive analytical approaches, often exceeding the dynamic range of currently available proteomic platforms. Moreover, and although saliva collection is non-invasive, sometimes it is not possible to have access to large volumes. Different CE-based approaches have been described in the study of salivary protein components. Conventional electrophoretic analysis of parotid saliva by SDS-PAGE is hampered by the fact that a number of proteins, such is the case of PRPs, are only poorly stained by currently used stains (as it was described in section 3.1). In this context, CE was suggested to offer advantages for the quantitation of the main components of parotid saliva, including α-amylase and a number of PRPs [92]. Analysis of salivary peptides, have also been reported, namely separation of histatins [93], which are a group of neutral to basic low molecular-weight proteins of human saliva with relevant antifungal properties. Moreover, substance P, a peptide present in saliva at trace levels, in the picomolar to nanomolar range, that functions as a neurotransmitter and/or as neuromodulator, has been analysed in this fluid by a CE method [94], circumventing the problems of detection of such small amounts. Another challenging approach has been presented in the case of secretory immunoglobulin A (sIgA) that was analysed in saliva by CE in association with immuno-fluorescence-labelling detection, which was considered an enormous advantage over radioimmunodiffusion or enzyme-linked immunosorbent assay (ELISA) usually used to assess the amounts of this salivary protein [95]. For proteome study, capillary isoelectric focusing has been used in combination with nano-reversed phase HPLC and mass spectrometry to profile and identify over a thousand salivary proteins [96]. The use of transient capillary isotachophoresis/capillary zone electrophoresis had been used for allowing the reduction of high-abundance proteins such as amylases, mucins, PRPs, and sIgA complex, what provides unparalleled advantages toward the identification of low-abundance proteins [97].

Saliva study based on the use of CE was not limited to the separation of proteins and peptides. Fluorescein-labeled amines were separated and detected in saliva using a microchip CE apparatus [98]. Additionally, some researchers have used this separation technique for the analysis of inorganic ions in saliva. The UV absorbing anions nitrate, nitrite and thiocyanate were determined in saliva by micellar electrokinetic capillary chromatography employing N-tetradecyl-N,Ndimethyl-3-ammonio-1-propanesulfonate (zwittergent-3-14) as a surfactant additive to the BGE [99]. Other slightly different micellar electrokinetic capillary chromatography procedures were optimized for these ions and are described elsewhere [e.g. [100–102]].

In salivary studies related to taste sensitivity and ingestive behaviour, the use of CE-based approaches may therefore offer advantages. Besides the already reported advantages for the study of saliva, some of the results obtained so far suggest the potential involvement of peptides in taste sensitivity and ingestive behaviour [e.g. [10]]. Hormones and peptides related to ingestive behaviour, known to be present in saliva are present in relatively low amounts (e.g. PYY, leptin and ghrelin) [12,103], and in this context their analysis can also be benefit from some of the advantages in CE. Nonetheless, this is an issue deserving further attention.

3.3.2. Electrophoresis in formalin-fixed and paraffin-embedded tissue samples

The fixation of tissues in formalin followed by embedding in paraffin is the standard tissue fixation and storage method adopted by most health institutions of pathology and histology departments, for research on biomarkers and molecular mechanisms of diseases. A great deal of information on proteins involved in many biological aspects is encased within these formalin-fixed paraffin-embedded (FFPE) tissue repositories, which offer an extensive resource for conducting retrospective and prospective studies.

The FFPE tissues are highly stable, even stored at room temperature, with cell or tissue structure being mainly preserved. The stability of proteins is achieved through cross-linking, with multiple reports showing that protein modifications, such as phosphorylation are maintained and can be determined years later by immunohistochemistry [104]. As such the FFPE tissues have the potential to constitute a valuable source of samples for proteome analysis using electrophoresis [105,106]. Protein extraction from FFPE tissue samples allows the generation of protein profiles by high-throughput techniques, providing an improvement over other traditional methods such as immunohistochemistry [107,108].

However, many questions remain in terms of quantification of the expression levels of proteins extracted from FFPE tissue samples and this approach has proven to be a daunting task with success varying according to the biomarker of interest and the tissue type [109,110]. In fact, the protein extraction efficiency is reduced by the extensive molecular crosslinking that occurs upon formalin fixation, with possible interferences in immunoreactivity or even in protein identification by mass spectrometry [111], imposing the development of efficient extraction methods [109,110,112,113]. The achievement of protein recovery with minimal losses in immunoreactivity is particularly relevant in studies needing immunoblotting. Several studies report methods of collecting and processing tissues directly from selected areas of histological sections using different tissue types [see extensive review in [113]], including salivary glands [114–116]. Most studies revealed comparability between protein extracts from FFPE and from unfixed fresh frozen tissue samples by different validation methods including SDS–PAGE and Western blotting [105,109,114,116,117].

The use of alternative fixatives, their advantages and disadvantages and potential for performing both morphological and molecular analyses on paraffin-embedded tissue sample, for proteomic applications and genome-wide expression analysis have been extensively investigated [114,117–120]. Another alternative for overcoming the problems posed by fixation with formalin passes through the development of robust long-term room temperature biospecimen tissue storage technologies that provide high quality nucleic acids or proteins [121].

Functional and morphological/histological changes of salivary glands are associated with various diet-related diseases, such as diabetes [122,123], hyperlipidemia [124], and obesity [125]. Moreover, the physical and chemical properties of foods affect salivary glands morphology and histology [126–129]. Although the regulation of the secretory activity of salivary glands has been almost exclusively attributed to the autonomic nervous system [130], recent evidences suggest that during variations in appetite and food intake the glands are probably also regulated by gastric (gastrin) and intestinal (cholecystokinin and melatonin) hormones [131–133]. Additionally to the peptide-hormones involved in short-term regulation of food-intake, the action of leptin in the salivary gland should not be discarded considering that the the presence of leptin receptors in salivary glands suggests in situ effects of this hormone [134,135]. As such, not only research in saliva fluid, but also in salivary glands, is valuable in studies related to ingestive behaviour.

Nonetheless, with optimized extraction methods, FFPE tissue samples can be a valuable source of protein, allowing reproducible and biologically relevant proteomic profiles for research on biomarkers and biological molecular mechanisms [109,110,112,136], with particular interest in systematic analysis of saliva and salivary gland feed-back regulation. However, despite the number of studies taking advantage of electrophoresis in FFPE tissue samples, currently, there is no consensus on the optimal protocol for protein extraction, neither accepted standards for

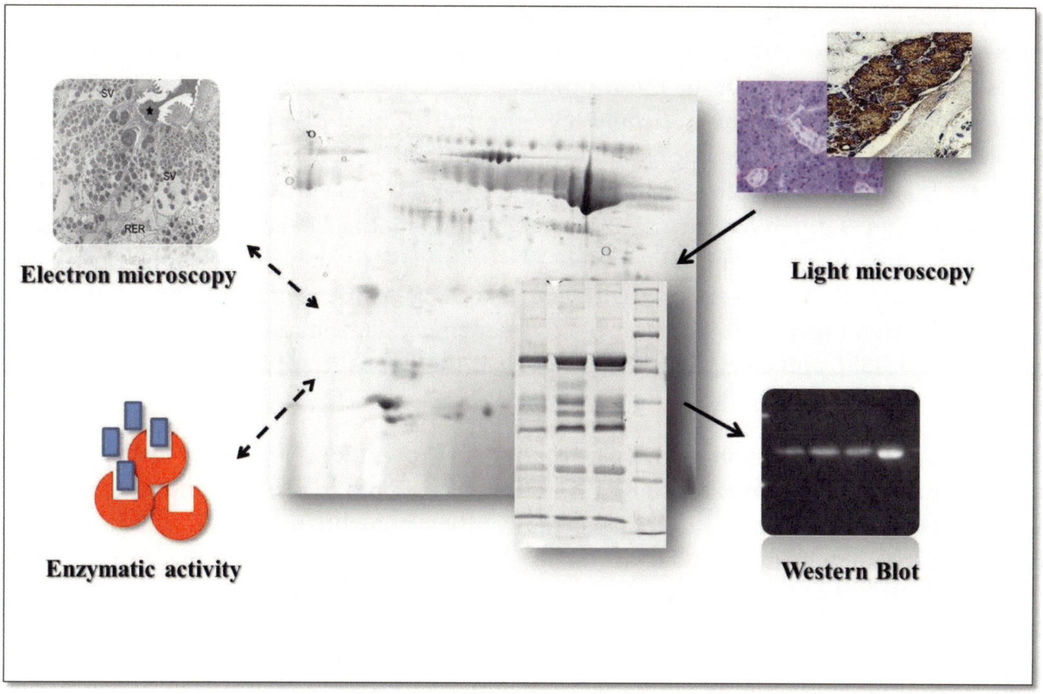

Figure 2. An integrative approach for studying saliva: techniques that can complement electrophoresis.

quantitative evaluation of the extracts. As such, further research is recommended to develop standardized methods ensuring quantitative and qualitative reproducibility in the protein recovery.

3.4. Complementarity of techniques — Contribution to a best case scenario

The potential of the different electrophoretic techniques, for the study of salivary secretion related to food perception and ingestive behaviour, alone or in combination, have been presented so far. However, the knowledge in this research area can be increased by combining different techniques that can give information which cannot be obtained through electrophoresis (Figure 2).

3.4.1. Enzymatic activity analysis

Some of the most abundant proteins in saliva are enzymes. These are the cases of: α-amylase, which, among several functions in saliva, begins the process of carbohydrate digestion in the mouth; and CA6, responsible for the maintenance of pH in oral cavity. Both of these enzymes are described to play a role in oral food perception and/or taste sensitivity, since they may contribute to changes in the concentration of simple sugars in the mouth and affect the viscosity of starch based products, in the case of α-amylase [137], or influence the conditions of the medium surrounding taste buds, in the case of CA6 [7].

The knowledge about the expression of salivary enzymes is important and can be achieved through electrophoretic techniques. However, proteome by its owns does not represent, necessarily, the functionality of these proteins. The complementarity between enzymatic activity data acquisition and electrophoretic protein profiles is also valuable for enzymes present in several different isoforms, being α-amylase and CA6 two examples [83]. In most of the cases the different isoforms are not all involved in the same physiological pathways and frequently express differences in activity and regulation motifs. Enzyme activity is dependent namely on the allosteric and fine-tuned phosphorylation/dephosphorylation regulation processes, thus reflecting an average of the activity of several isoenzymes present in the sample in a specific environment. For example, α-amylase activity depends on the presence of calcium and chloride [138], and changes in the amounts of these ions result in variations in enzyme activity.

Moreover, although two-dimensional electrophoresis allows the separation, visualization, and even the quantification, of the different isoforms (based both in different molecular masses and *pIs*), as well the discrimination of isoforms resulting from different PTMs, the major drawback of this technique resides in the fact that, by being an at least partially denaturing method, it does not allow the identification of the native structure of an enzyme but its subunits only. In this context, enzyme activity analysis offers functional information that is complementary to proteome analysis by electrophoresis. The acquisition of electrophoretic profiles and enzymatic activity data, simultaneously, can greatly amplify the physiological interpretation of results, being valuable in the analysis of the interaction saliva-food perception. Previous results, from studies in obesity, obtained in our laboratory, present evidences of the meaning of such complementarity, showing that minor changes in protein expression, may be associated with significant differences in enzymatic activity [139]. For instances, a higher α-amylase activity was not accompanied by changes in the expression of this protein, evidenced by western blot analysis. Despite the unchanged expression, it was hypothesised that this increase in enzymatic activity could be related to changes in sweet food perception and acceptance, reported by other authors for obese individuals [140]. Moreover, the evaluation of enzymatic activity has the advantages of being relatively inexpensive, in comparison with some electrophoretic techniques and of offering the possibility of being almost fully automated. The most common methodologies applied are dependent on spectrophotometry or fluorimetry techniques. The use of the multi-well plate readers allows the analysis of a high number of samples, with minimal reagent consumption.

3.4.2. Microscopy techniques

Changes in salivary proteome related to food perception and ingestion can be better understood if complemented with microscopy imaging techniques. These may improve the knowledge about the secretory mechanisms and cell type contributing for the secretion of salivary proteins, its expression and localization, i.e. information not possible to obtain through electrophoretic salivary protein profiles alone. Different studies have used different microscopy based techniques in salivary gland research. Some examples are: the observation of subcellular distribution of lysozyme in the mouse major salivary glands, the intragranular compartmentation of this secretion enzyme and its relationship with α-amylase using light and electron microscopy [141]; the study of contraction of myoepithelial cells in the human submandibular gland using confocal microscopy [142]. Moreover, the combination of electrophoretic and immunohistochemical techniques have been used to evaluate the distribution and expression of several proteins, such as CA6 in minor and major salivary glands of humans [25,143], leptin and the functional leptin receptor in major salivary glands of humans [135], muscarinic receptor subtypes in salivary glands of rats, sheep and man [144], aquaporin water channels in rat major salivary glands [145,146], and cytoskeleton proteins in rat parotid glands [147].

The salivary glands are a good model for the study of exocrine secretion. The proteins that will be secreted are synthesized in the endoplasmic reticulum and transported through the Golgi apparatus to the trans-Golgi network where they are stored in secretory granules (SCGs) after which they are released into the cytoplasm and transported to the periphery of the cell [148]. Electron microscopy studies have shown that the secretory cells of salivary glands are ultrastructurally equipped to produce and store large amounts of secretory proteins in secretory granules (SGs) [149]. The volume and number of SGs per cell are a morphological evidence of the storage capacity of the secretory cells. Regarding the secretory cell activity, the number and development of organelles, as well as membrane cell specializations, are ultrastructural traits associated with the mechanism of secretory production. A recent study reported differences in secretory activity and volume of saliva secreted by goat parotid glands (PGs) between liquid and solid diets: milk-suckling kids (MSKs) and diet-fed goats (DFGs) [150]. Such differences were correlated with the ultrastruture of the secretory cells. In PGs of DFGs a well-developed system of membrane specializations was identified, which was not prominent in PGs of MSKs. Such a characteristic is typical of cells in phase of intense secretory activity [151]. Moreover, the presence of well-developed smooth muscular cells around the secretory endpieces of the PGs of DFGs, but not in MSKs, is a morphological evidence of the release of large amounts of secretion into the acinar lumen. This is also supported by the low number of SGs observed in the apical cytoplasm of the secretory cells of DFGs as compared with that in MSKs.

The ultrastructure of the salivary glands can also reflect the chemical composition of the secretion and the subcellular localization of the salivary proteins of interest. Secretory cells of the three major types of salivary glands possess SGs with variable appearance and density under electron microscopy. Presumably, these morphological features of the SGs are indicative of constituent differences in secretory protein content [152]. Immunohistochemistry coupled to electron microscopy proved also to be useful by providing an increased knowledge of localization and distribution of the different salivary proteins in the parenchyma of major and minor salivary glands [e.g. [153,154]]. From the different studies, evidences arrived that different proteins can be differently distributed among secretory granules and that different mechanisms can be involved in the packaging of the different salivary proteins. In conclusion, the electron microscopy coupled to the cytochemistry may provide valuable knowledge on cell biology of the salivary glands. This information is a basis to know the morphological and functional diversity of these glands, which allows understanding the modes of production and secretion, as well as the types of secretion correlated with ingestive behavior.

4. Concluding remarks

Saliva study received considerable attention in the last years, particularly due to the potential of this fluid as a non-invasive source of biomarkers. Despite the focus had been put mainly in pathological conditions, it is known that salivary proteins can be useful for the understanding of physiological mechanisms. Research in the area of ingestive behaviour can be improved with the study of saliva. Salivary proteins, besides participating, by different manners, in food perception, may also be markers of taste sensitivity and/or food choices with potential application in the management of dietary programs.

Despite the existence of limitations, electrophoresis continues to be an essential tool in the study of salivary proteome. It constitutes the basis for the separation of several different components, allowing a summary characterization and also providing a purification step prior the application of more selective and commonly more expensive methods. Several human and animal salivary proteins, involved in taste perception and food choices, have been separated through electrophoretic techniques, for further identification. Nevertheless, enhanced methodologies for sample fractionation and processing might be useful to circumvent some of the limitations in the study of this fluid by electrophoresis. The possibility of using electrophoresis in samples obtained from formalin-fixed paraffin embedded tissues, for different purposes, has been presented as a great advantage for the study of long-term stored samples. However, this approach has rarely been applied to non-malignant salivary tissue samples, being mainly used in quantitation of protein biomarkers in clinical tissue specimens. Non-electrophoretic techniques, such as microscopy and/or studies of enzymatic activity, can greatly add in saliva research related to ingestive behaviour. Knowledge in secretion induced by dietary compounds can benefit from structural and ultra-structural information. Moreover, protein expression is not necessarily concordant with enzymatic activity.

In conclusion, new and improved approaches will be valuable to cope with the challenges in understanding the influence of saliva in food perception and ingestive behaviour. The study of saliva appears to be promising in biomarker discovery contributing for the understanding of food choices, hence having major impact in human and animal health, and major fundamental relevance for animal production. As such further integrated and systematic analysis of saliva, combining old and new approaches, will profit from complementarity of different methodologies.

Acknowledgements

This chapter is funded by FEDER Funds through the Operational Programme for Competitiveness Factors-COMPETE and National Funds through FCT-Foundation for Science and

Technology under the Strategic Projects PEst-C/AGR/UI0115/2011, PEst-OE/AGR/UI0115/2014, and PEst-C/SAU/LA0001/2011. Authors acknowledge also the financial support from the Portuguese Science Foundation (FCT) in the form of Elsa Lamy FCT investigator contract IF/01778/2013. The Portuguese Science Foundation (FCT) played no role in the development of the present work or upon its submission for publication.

Author details

Elsa Lamy[1*], Martine Morzel[2], Lénia Rodrigues[1], Cristina Pinheiro[3], Ana Rodrigues Costa[4], Célia Miguel Antunes[5], Orlando Lopes[6] and Fernando Capela e Silva[6]

*Address all correspondence to: ecsl@uevora.pt

1 Instituto de Ciências Agrárias e Ambientais Mediterrânicas (ICAAM), Instituto de Investigação e Formação Avançada, Universidade de Évora, Évora, Portugal

2 CSGA-Centre des Sciences du Goût et de l'Alimentation, INRA, CNRS, Université de Bourgogne, Dijon, France

3 Departamento de Zootecnia, Escola de Ciências e Tecnologias, Instituto de Ciências Agrárias e Ambientais Mediterrânicas (ICAAM), Instituto de Investigação e Formação Avançada, Universidade de Évora, Évora, Portugal

4 Departamento de Química, Escola de Ciências e Tecnologias, Instituto de Ciências Agrárias e Ambientais Mediterrânicas (ICAAM), Instituto de Investigação e Formação Avançada, Universidade de Évora, Évora, Portugal

5 Departamento de Química, Escola de Ciências e Tecnologias, Universidade de Évora, Évora, Portugal e Centro de Neurociências e Biologia Celular (CNC), Universidade de Coimbra, Coimbra, Portugal

6 Departamento de Biologia, Escola de Ciências e Tecnologias, Instituto de Ciências Agrárias e Ambientais Mediterrânicas (ICAAM), Instituto de Investigação e Formação Avançada, Universidade de Évora, Évora, Portugal

References

[1] Ruhl S. The scientific exploration of saliva in the post-proteomic era: from database back to basic function. Expert Rev Proteomics. 2012 Jan;9(1):85–96.

[2] Lamy E, da Costa G, Santos R, Capela E Silva F, Potes J, Pereira A, et al. Sheep and goat saliva proteome analysis: a useful tool for ingestive behavior research? Physiol Behav. 2009 Oct 19;98(4):393–401.

[3] Lamy E, Graça G, da Costa G, Franco C, E Silva FC, Baptista ES, et al. Changes in mouse whole saliva soluble proteome induced by tannin-enriched diet. Proteome Sci. 2010 Jan;8:65.

[4] Lamy E, da Costa G, Santos R, Capela e Silva F, Potes J, Pereira A, et al. Effect of condensed tannin ingestion in sheep and goat parotid saliva proteome. J Anim Physiol Anim Nutr (Berl). 2011 Jun;95(3):304–12.

[5] Morzel M, Chabanet C, Schwartz C, Lucchi G, Ducoroy P, Nicklaus S. Salivary protein profiles are linked to bitter taste acceptance in infants. Eur J Pediatr. 2014 May; 173(5):575–82.

[6] Bajec MR, Pickering GJ. Astringency: mechanisms and perception. Crit Rev Food Sci Nutr. 2008 Oct;48(9):858–75.

[7] Henkin RI, Martin BM, Agarwal RP. Decreased parotid saliva gustin/carbonic anhydrase VI secretion: an enzyme disorder manifested by gustatory and olfactory dysfunction. Am J Med Sci. 1999 Dec;318(6):380–91.

[8] Neyraud E, Palicki O, Schwartz C, Nicklaus S, Feron G. Variability of human saliva composition: possible relationships with fat perception and liking. Arch Oral Biol. 2012 May;57(5):556–66.

[9] Cabras T, Melis M, Castagnola M, Padiglia A, Tepper BJ, Messana I, et al. Responsiveness to 6-n-propylthiouracil (PROP) is associated with salivary levels of two specific basic proline-rich proteins in humans. PLoS One. 2012 Jan;7(2):e30962.

[10] Dsamou M, Palicki O, Septier C, Chabanet C, Lucchi G, Ducoroy P, et al. Salivary protein profiles and sensitivity to the bitter taste of caffeine. Chem Senses. 2012 Jan; 37(1):87–95.

[11] Gröschl M, Rauh M, Wagner R, Neuhuber W, Metzler M, Tamgüney G, et al. Identification of leptin in human saliva. J Clin Endocrinol Metab. 2001 Nov;86(11):5234–9.

[12] Gröschl M, Topf HG, Bohlender J, Zenk J, Klussmann S, Dötsch J, et al. Identification of ghrelin in human saliva: production by the salivary glands and potential role in proliferation of oral keratinocytes. Clin Chem. 2005 Jun;51(6):997–1006.

[13] Lamy E, da Costa G, Santos R, Capela E Silva F, Potes J, Pereira A, et al. Sheep and goat saliva proteome analysis: a useful tool for ingestive behavior research? Physiol Behav. 2009 Oct 19;98(4):393–401.

[14] Petersen JR, Okorodudu AO, Mohammad A, Payne DA. Capillary electrophoresis and its application in the clinical laboratory. Clin Chim Acta. 2003 Apr;330(1-2):1–30.

[15] Fowler CB, O'Leary TJ, Mason JT. Toward improving the proteomic analysis of formalin-fixed, paraffin-embedded tissue. Expert Rev Proteomics. 2013 Aug;10(4):389–400.

[16] Simons PJ, Kummer JA, Luiken JJFP, Boon L. Apical CD36 immunolocalization in human and porcine taste buds from circumvallate and foliate papillae. Acta Histochem. 2011 Dec;113(8):839–43.

[17] Galindo MM, Voigt N, Stein J, van Lengerich J, Raguse J-D, Hofmann T, et al. G protein-coupled receptors in human fat taste perception. Chem Senses. 2012 Feb;37(2): 123–39.

[18] Schmale H, Holtgreve-Grez H, Christiansen H. Possible role for salivary gland protein in taste reception indicated by homology to lipophilic-ligand carrier proteins. Nature. 1990 Jan 25;343(6256):366–9.

[19] Li XJ, Snyder SH. Molecular cloning of Ebnerin, a von Ebner's gland protein associated with taste buds. J Biol Chem. 1995 Jul 28;270(30):17674–9.

[20] Bläker M, Kock K, Ahlers C, Buck F, Schmale H. Molecular cloning of human von Ebner's gland protein, a member of the lipocalin superfamily highly expressed in lingual salivary glands. Biochim Biophys Acta. 1993 Feb 20;1172(1-2):131–7.

[21] Spielman AI, D'Abundo S, Field RB, Schmale H. Protein analysis of human von Ebner saliva and a method for its collection from the foliate papillae. J Dent Res. 1993 Sep;72(9):1331–5.

[22] Poette J, Mekoué J, Neyraud E, Berdeaux O, Renault A, Guichard E, et al. Fat sensitivity in humans: oleic acid detection threshold is linked to saliva composition and oral volume. Flavour Fragr J. 2014 Jan 17;29(1):39–49.

[23] Shatzman AR, Henkin RI. Gustin concentration changes relative to salivary zinc and taste in humans. Proc Natl Acad Sci U S A. 1981 Jun;78(6):3867–71.

[24] Henkin RI, Martin BM, Agarwal RP. Efficacy of exogenous oral zinc in treatment of patients with carbonic anhydrase VI deficiency. Am J Med Sci. 1999 Dec;318(6):392–405.

[25] Leinonen J, Parkkila S, Kaunisto K, Koivunen P, Rajaniemi H. Secretion of carbonic anhydrase isoenzyme VI (CA VI) from human and rat lingual serous von Ebner's glands. J Histochem Cytochem. 2001 May;49(5):657–62.

[26] Padiglia A, Zonza A, Atzori E, Chillotti C, Calò C, Tepper BJ, et al. Sensitivity to 6-n-propylthiouracil is associated with gustin (carbonic anhydrase VI) gene polymorphism, salivary zinc, and body mass index in humans. Am J Clin Nutr. 2010 Sep; 92(3):539–45.

[27] Feeney EL, Hayes JE. Exploring associations between taste perception, oral anatomy and polymorphisms in the carbonic anhydrase (gustin) gene CA6. Physiol Behav. 2014 Apr 10;128:148–54.

[28] Watanabe M, Asatsuma M, Ikui A, Ikeda M, Yamada Y, Nomura S, et al. Measurements of several metallic elements and matrix metalloproteinases (MMPs) in saliva from patients with taste disorder. Chem Senses. 2005 Feb;30(2):121–5.

[29] Morris-Wiman J, Sego R, Brinkley L, Dolce C. The effects of sialoadenectomy and exogenous EGF on taste bud morphology and maintenance. Chem Senses. 2000 Feb; 25(1):9–19.

[30] Wada, O. H., Yamamori, T., Shimazaki, N., Ishibashi, K. & Seino K. Function of salivary histatin 5 for bitter substances. 88th General Session and Exhibition of the IADR (International Association for Dental Research). 2010. p. MM–598.

[31] Melis M, Aragoni MC, Arca M, Cabras T, Caltagirone C, Castagnola M, et al. Marked increase in PROP taste responsiveness following oral supplementation with selected salivary proteins or their related free amino acids. PLoS One. 2013 Jan;8(3):e59810.

[32] Mounayar R, Morzel M, Brignot H, Tremblay-Franco M, Canlet C, Lucchi G, et al. Salivary markers of taste sensitivity to oleic acid: a combined proteomics and metabolomics approach. Metabolomics. 2013 Oct 27;10(4):688–96.

[33] Matsuo R. Role of saliva in the maintenance of taste sensitivity. Crit Rev Oral Biol Med. 2000 Jan;11(2):216–29.

[34] ASTM. Annual book ofASTMstandards. Philadelphia: American Society for Testing and Materials; 2004.

[35] Lesschaeve I, Noble AC. Polyphenols: factors influencing their sensory properties and their effects on food and beverage preferences. Am J Clin Nutr. 2005 Jan;81(1 Suppl):330S–335S.

[36] Lee CA, Vickers ZM. Astringency of foods may not be directly related to salivary lubricity. J Food Sci. 2012 Sep;77(9):S302–6.

[37] Mehansho H, Butler LG, Carlson DM. Dietary tannins and salivary proline-rich proteins: interactions, induction, and defense mechanisms. Annu Rev Nutr. 1987 Jan; 7:423–40.

[38] C. T. Robbins AEH. Variation in mammalian physiological responses to a condensed tannin and its ecological implications. J Mammal. 1991;72:480 – 486.

[39] De Wijk Ra, Prinz Jf. Mechanisms Underlying the Role of Friction in Oral Texture. J Texture Stud. 2006 Aug;37(4):413–27.

[40] Rossetti D, Yakubov GE, Stokes JR, Williamson A-M, Fuller GG. Interaction of human whole saliva and astringent dietary compounds investigated by interfacial shear rheology. Food Hydrocoll. 2008 Aug;22(6):1068–78.

[41] Sarni-Manchado P, Canals-Bosch J-M, Mazerolles G, Cheynier V. Influence of the glycosylation of human salivary proline-rich proteins on their interactions with condensed tannins. J Agric Food Chem. 2008 Oct 22;56(20):9563–9.

[42] Schwarz B, Hofmann T. Is there a direct relationship between oral astringency and human salivary protein binding? Eur Food Res Technol. 2008 May 27;227(6):1693–8.

[43] Nayak A, Carpenter GH. A physiological model of tea-induced astringency. Physiol Behav. 2008 Oct 20;95(3):290–4.

[44] Williamson MP. The structure and function of proline-rich regions in proteins. Biochem J. 1994 Jan 15;297 (Pt 2:249–60.

[45] Yan Q, Bennick A. Identification of histatins as tannin-binding proteins in human saliva. Biochem J. 1995 Oct 1;311 (Pt 1:341–7.

[46] Soares S, Mateus N, de Freitas V. Interaction of different classes of salivary proteins with food tannins. Food Res Int. 2012 Dec;49(2):807–13.

[47] Amerongen AVN, Veerman ECI. Saliva--the defender of the oral cavity. Oral Dis. 2002 Jan;8(1):12–22.

[48] Canon F, Giuliani A, Paté F, Sarni-Manchado P. Ability of a salivary intrinsically unstructured protein to bind different tannin targets revealed by mass spectrometry. Anal Bioanal Chem. 2010 Sep;398(2):815–22.

[49] Jöbstl E, O'Connell J, Fairclough JPA, Williamson MP. Molecular model for astringency produced by polyphenol/protein interactions. Biomacromolecules. 5(3):942–9.

[50] Mehansho H, Clements S, Sheares BT, Smith S, Carlson DM. Induction of proline-rich glycoprotein synthesis in mouse salivary glands by isoproterenol and by tannins. J Biol Chem. 1985 Apr 10;260(7):4418–23.

[51] Chan M, Bennick A. Proteolytic processing of a human salivary proline-rich protein precursor by proprotein convertases. Eur J Biochem. 2001 Jun;268(12):3423–31.

[52] Asquith TN, Uhlig J, Mehansho H, Putman L, Carlson DM, Butler L. Binding of condensed tannins to salivary proline-rich glycoproteins: the role of carbohydrate. J Agric Food Chem. American Chemical Society; 1987 May;35(3):331–4.

[53] Boze H, Marlin T, Durand D, Pérez J, Vernhet A, Canon F, et al. Proline-rich salivary proteins have extended conformations. Biophys J. 2010 Jul 21;99(2):656–65.

[54] Gambuti A, Rinaldi A, Pessina R, Moio L. Evaluation of aglianico grape skin and seed polyphenol astringency by SDS–PAGE electrophoresis of salivary proteins after the binding reaction. Food Chem. 2006 Aug;97(4):614–20.

[55] Al-Tarawneh SK, Border MB, Dibble CF, Bencharit S. Defining salivary biomarkers using mass spectrometry-based proteomics: a systematic review. OMICS. 2011 Jun; 15(6):353–61.

[56] Siqueira WL, Dawes C. The salivary proteome: challenges and perspectives. Proteomics Clin Appl. 2011 Dec;5(11-12):575–9.

[57] Dawes C. Circadian rhythms in human salivary flow rate and composition. J Physiol. 1972 Feb;220(3):529–45.

[58] Inoue H, Ono K, Masuda W, Morimoto Y, Tanaka T, Yokota M, et al. Gender difference in unstimulated whole saliva flow rate and salivary gland sizes. Arch Oral Biol. 2006 Dec;51(12):1055–60.

[59] Scully C. Drug effects on salivary glands: dry mouth. Oral Dis. 2003 Jul;9(4):165–76.

[60] Khaustova S, Shkurnikov M, Tonevitsky E, Artyushenko V, Tonevitsky A. Noninvasive biochemical monitoring of physiological stress by Fourier transform infrared saliva spectroscopy. Analyst. 2010;135:3183–92.

[61] Humphrey SP, Williamson RT. A review of saliva: normal composition, flow, and function. J Prosthet Dent. 2001 Feb;85(2):162–9.

[62] Navazesh M. Methods for collecting saliva. Ann N Y Acad Sci. 1993 Sep 20;694:72–7.

[63] Lamy E, Costa AR, Antunes CM, Vitorino R, Amado F. Protein Electrophoresis in Saliva Study. In: Ghowsi K, editor. Electrophoresis. 2012.

[64] Xiao H, Wong DTW. Method development for proteome stabilization in human saliva. Anal Chim Acta. 2012 Apr 13;722:63–9.

[65] Thomadaki K, Helmerhorst EJ, Tian N, Sun X, Siqueira WL, Walt DR, et al. Whole-saliva proteolysis and its impact on salivary diagnostics. J Dent Res. 2011 Nov;90(11): 1325–30.

[66] Schipper R, Loof A, de Groot J, Harthoorn L, Dransfield E, van Heerde W. SELDI-TOF-MS of saliva: methodology and pre-treatment effects. J Chromatogr B Analyt Technol Biomed Life Sci. 2007 Feb 15;847(1):45–53.

[67] Presser E, Simuyandi M, Brown J. The effects of storage time and temperature on recovery of salivary secretory immunoglobulin A. Am J Hum Biol. 26(3):417–20.

[68] Deutsch O, Fleissig Y, Zaks B, Krief G, Aframian DJ, Palmon A. An approach to remove alpha amylase for proteomic analysis of low abundance biomarkers in human saliva. Electrophoresis. 2008 Nov;29(20):4150–7.

[69] Krief G, Deutsch O, Zaks B, Wong DT, Aframian DJ, Palmon A. Comparison of diverse affinity based high-abundance protein depletion strategies for improved biomarker discovery in oral fluids. J Proteomics. 2012 Jul 16;75(13):4165–75.

[70] Patton WF. Detection technologies in proteome analysis. J Chromatogr B Analyt Technol Biomed Life Sci. 2002 May 5;771(1-2):3–31.

[71] Beeley JA, Newman F, Wilson PH, Shimmin IC. Sodium dodecyl sulphate-polyacrylamide gel electrophoresis of human parotid salivary proteins: comparison of dansy-

lation, coomassie blue R-250 and silver detection methods. Electrophoresis. 1996 Mar;17(3):505–6.

[72] Steinberg TH. Protein gel staining methods: an introduction and overview. Methods Enzymol. 2009 Jan;463:541–63.

[73] Miller I, Crawford J, Gianazza E. Protein stains for proteomic applications: which, when, why? Proteomics. 2006 Oct;6(20):5385–408.

[74] Sondej M, Denny PA, Xie Y, Ramachandran P, Si Y, Takashima J, et al. Glycoprofiling of the Human Salivary Proteome. Clin Proteomics. 2009 Mar 1;5(1):52–68.

[75] Costa AR, Real J, Antunes CM, Cruz-Morais J. A new approach for determination of Na,K-ATPase activity: application to intact pancreatic β-cells. In Vitro Cell Dev Biol Anim. 2010 Jan;46(1):7–10.

[76] Nakamura K, Tanaka T, Kuwahara A, Takeo K. Microassay for proteins on nitrocellulose filter using protein dye-staining procedure. Anal Biochem. 1985 Aug 1;148(2): 311–9.

[77] Salinovich O, Montelaro RC. Comparison of glycoproteins by two-dimensional mapping of glycosylated peptides. Anal Biochem. 1986 Aug 15;157(1):19–27.

[78] Mau M, Kaiser TM, Südekum K-H. Carbonic anhydrase II is secreted from bovine parotid glands. Histol Histopathol. 2010 Mar;25(3):321–9.

[79] Gorr S-U, Venkatesh SG, Darling DS. Parotid secretory granules: crossroads of secretory pathways and protein storage. J Dent Res. 2005 Jun;84(6):500–9.

[80] Lamy E, da Costa G, Capela e Silva F, Potes J, Coelho AV, Baptista ES. Comparison of electrophoretic protein profiles from sheep and goat parotid saliva. J Chem Ecol. 2008 Mar;34(3):388–97.

[81] Lamy E, Mau M. Saliva proteomics as an emerging, non-invasive tool to study livestock physiology, nutrition and diseases. J Proteomics. 2012 Jul 19;75(14):4251–8.

[82] Al-Tarawneh SK, Border MB, Dibble CF, Bencharit S. Defining salivary biomarkers using mass spectrometry-based proteomics: a systematic review. OMICS. 2011 Jun; 15(6):353–61.

[83] Hirtz C, Chevalier F, Centeno D, Rofidal V, Egea J-C, Rossignol M, et al. MS characterization of multiple forms of alpha-amylase in human saliva. Proteomics. 2005 Nov;5(17):4597–607.

[84] Neyraud E, Sayd T, Morzel M, Dransfield E. Proteomic analysis of human whole and parotid salivas following stimulation by different tastes. J Proteome Res. 2006 Sep; 5(9):2474–80.

[85] Morzel M, Chabanet C, Schwartz C, Lucchi G, Ducoroy P, Nicklaus S. Salivary protein profiles are linked to bitter taste acceptance in infants. Eur J Pediatr. 2014 May; 173(5):575–82.

[86] Lamy E, Graça G, da Costa G, Franco C, E Silva FC, Baptista ES, et al. Changes in mouse whole saliva soluble proteome induced by tannin-enriched diet. Proteome Sci. 2010 Jan;8:65.

[87] Lamy E, da Costa G, Santos R, Capela e Silva F, Potes J, Pereira A, et al. Effect of condensed tannin ingestion in sheep and goat parotid saliva proteome. J Anim Physiol Anim Nutr (Berl). 2011 Jun;95(3):304–12.

[88] Dinnella C, Recchia A, Fia G, Bertuccioli M, Monteleone E. Saliva characteristics and individual sensitivity to phenolic astringent stimuli. Chem Senses. 2009 May;34(4): 295–304.

[89] Dinnella C, Recchia A, Vincenzi S, Tuorila H, Monteleone E. Temporary modification of salivary protein profile and individual responses to repeated phenolic astringent stimuli. Chem Senses. 2010 Jan;35(1):75–85.

[90] Salas-Solano O, Kennel B, Park SS, Roby K, Sosic Z, Boumajny B, et al. Robustness of iCIEF methodology for the analysis of monoclonal antibodies: an interlaboratory study. J Sep Sci. 2012 Nov;35(22):3124–9.

[91] Hancu G, Simon B, Rusu A, Mircia E, Gyéresi A. Principles of micellar electrokinetic capillary chromatography applied in pharmaceutical analysis. Adv Pharm Bull. 2013 Jan;3(1):1–8.

[92] Dieijen-Visser Ymch and MP van. Capillary zone electrophoresis as a tool to detect proteins in body fluids: reproducibility, comparison with conventional methods and a review of the literature. Ned Tijdschr Klin Chem. 2000;25:219–29.

[93] Lal K, Xu L, Colburn J, Hong AL, Pollock JJ. The use of capillary electrophoresis to identify cationic proteins in human parotid saliva. Arch Oral Biol. 1992 Jan;37(1):7–13.

[94] Frerichs VA, Herrmann JK, Aguirre A, Colón LA. Trace level determination of substance P using capillary electrophoresis and laser-induced fluorescence. Microchem J. 2004 Oct;78(2):135–42.

[95] Liu C-M, Tung K-H, Chang T-H, Chien C-C, Yen M-H. Analysis of secretory immunoglobulin A in human saliva by laser-induced fluorescence capillary electrophoresis. J Chromatogr B Analyt Technol Biomed Life Sci. 2003 Jul 5;791(1-2):315–21.

[96] Guo T, Rudnick PA, Wang W, Lee CS, Devoe DL, Balgley BM. Characterization of the human salivary proteome by capillary isoelectric focusing/nanoreversed-phase liquid chromatography coupled with ESI-tandem MS. J Proteome Res. 2006 Jun;5(6): 1469–78.

[97] Fang X, Yang L, Wang W, Song T, Lee CS, DeVoe DL, et al. Comparison of electrokinetics-based multidimensional separations coupled with electrospray ionization-tandem mass spectrometry for characterization of human salivary proteins. Anal Chem. 2007 Aug 1;79(15):5785–92.

[98] Tanaka Y, Naruishi N, Nakayama Y, Higashi T, Wakida S. Development of an analytical method using microchip capillary electrophoresis for the measurement of fluorescein-labeled salivary components in response to exercise stress. J Chromatogr A. 2006 Mar 24;1109(2):132–7.

[99] Mori M, Hu W, Fritz JS, Tsue H, Kaneta T, Tanaka S. Determination of inorganic anions in human saliva by zwitterionic micellar capillary electrophoresis. Fresenius J Anal Chem. 2001 Jun;370(4):429–33.

[100] Tanaka Y, Naruishi N, Fukuya H, Sakata J, Saito K, Wakida S-I. Simultaneous determination of nitrite, nitrate, thiocyanate and uric acid in human saliva by capillary zone electrophoresis and its application to the study of daily variations. J Chromatogr A. 2004 Oct 8;1051(1-2):193–7.

[101] Vitali L, Fávere VT, Micke GA. A new method to determine biological sample volume by short end multiple injection capillary electrophoresis: application in determination of nitrate and thiocyanate in human saliva. J Chromatogr A. 2011 Apr 22;1218(16):2327–33.

[102] Ren H, Qiu H, Liang X, Wang X, Jiang S. Determination of inorganic anions in saliva by electroosmotic flow controlled counterflow isotachophoretic stacking under field-amplified sample injection. J Chromatogr B Analyt Technol Biomed Life Sci. 2013 Sep 15;935:75–9.

[103] Acosta A, Hurtado MD, Gorbatyuk O, La Sala M, Duncan D, Aslanidi G, et al. Salivary PYY: a putative bypass to satiety. PLoS One. Public Library of Science; 2011 Jan 10;6(10):e26137.

[104] Lim MS, Elenitoba-Johnson KSJ. Proteomics in pathology research. Lab Invest. 2004 Oct;84(10):1227–44.

[105] Krebs B, Kohlmannsperger V, Nölting S, Schmalzbauer R, Kretzschmar HA. A method to perform Western blots of microscopic areas of histological sections. J Histochem Cytochem. 2006 May;54(5):559–65.

[106] Zhu L, Tangrea MA, Mukherjee S, Emmert-Buck MR. Layered electrophoretic transfer-A method for pre-analytic processing of histological sections. Proteomics. 2011 Mar;11(5):883–9.

[107] Aerni H-R, Cornett DS, Caprioli RM. High-throughput profiling of formalin-fixed paraffin-embedded tissue using parallel electrophoresis and matrix-assisted laser desorption ionization mass spectrometry. Anal Chem. 2009 Sep 1;81(17):7490–5.

[108] Matsuda KM, Chung J-Y, Hewitt SM. Histo-proteomic profiling of formalin-fixed, paraffin-embedded tissue. Expert Rev Proteomics. 2010 Apr;7(2):227–37.

[109] Gräntzdörffer I, Yumlu S, Gioeva Z, von Wasielewski R, Ebert MPA, Röcken C. Comparison of different tissue sampling methods for protein extraction from formalin-fixed and paraffin-embedded tissue specimens. Exp Mol Pathol. 2010 Feb;88(1):190–6.

[110] Klopfleisch R, Weiss ATA, Gruber AD. Excavation of a buried treasure--DNA, mRNA, miRNA and protein analysis in formalin fixed, paraffin embedded tissues. Histol Histopathol. 2011 Jun;26(6):797–810.

[111] Crockett DK, Lin Z, Vaughn CP, Lim MS, Elenitoba-Johnson KSJ. Identification of proteins from formalin-fixed paraffin-embedded cells by LC-MS/MS. Lab Invest. 2005 Nov;85(11):1405–15.

[112] Guo H, Liu W, Ju Z, Tamboli P, Jonasch E, Mills GB, et al. An efficient procedure for protein extraction from formalin-fixed, paraffin-embedded tissues for reverse phase protein arrays. Proteome Sci. 2012 Jan;10(1):56.

[113] Maes E, Broeckx V, Mertens I, Sagaert X, Prenen H, Landuyt B, et al. Analysis of the formalin-fixed paraffin-embedded tissue proteome: pitfalls, challenges, and future prospectives. Amino Acids. 2013 Aug;45(2):205–18.

[114] Gündisch S, Schott C, Wolff C, Tran K, Beese C, Viertler C, et al. The PAXgene(®) tissue system preserves phosphoproteins in human tissue specimens and enables comprehensive protein biomarker research. PLoS One. 2013 Jan;8(3):e60638.

[115] Kokkat TJ, Patel MS, McGarvey D, LiVolsi VA, Baloch ZW. Archived formalin-fixed paraffin-embedded (FFPE) blocks: A valuable underexploited resource for extraction of DNA, RNA, and protein. Biopreserv Biobank. 2013 Apr;11(2):101–6.

[116] Celes M, Soave D, Freitas AC, Ramos S, Ribeiro-Silva A. Immunoblotting analysis of formalin-fixed and paraffin-embedded malignant tumor samples (1048.16). FASEB J. 2014 Apr 1;28(1_Supplement):1048.16–.

[117] Bellet V, Boissière F, Bibeau F, Desmetz C, Berthe ML, Rochaix P, et al. Proteomic analysis of RCL2 paraffin-embedded tissues. J Cell Mol Med. 2008 Oct;12(5B):2027–36.

[118] Ahram M, Flaig MJ, Gillespie JW, Duray PH, Linehan WM, Ornstein DK, et al. Evaluation of ethanol-fixed, paraffin-embedded tissues for proteomic applications. Proteomics. 2003 Apr;3(4):413–21.

[119] Denouël A, Boissière-Michot F, Rochaix P, Bibeau F, Boulle N. An alternative fixative to formalin fixation for molecular applications: the RCL2(®)-CS100 approach. Methods Mol Biol. 2011 Jan;724:297–307.

[120] Boissière-Michot F, Denouël A, Boulle N, Guillaume C, Orsetti B, Lopez-Crapez E, et al. The non-crosslinking fixative RCL2®-CS100 is compatible with both pathology diagnosis and molecular analyses. Pathol Oncol Res. 2013 Jan;19(1):41–53.

[121] Lou JJ, Mirsadraei L, Sanchez DE, Wilson RW, Shabihkhani M, Lucey GM, et al. A review of room temperature storage of biospecimen tissue and nucleic acids for anatomic pathology laboratories and biorepositories. Clin Biochem. 2014 Mar;47(4-5): 267–73.

[122] Carda C, Mosquera-Lloreda N, Salom L, Gomez de Ferraris ME, Peydró A. Structural and functional salivary disorders in type 2 diabetic patients. Med Oral Patol Oral Cir Bucal. 2006 Jul;11(4):E309–14.

[123] Lilliu MA, Solinas P, Cossu M, Puxeddu R, Loy F, Isola R, et al. Diabetes causes morphological changes in human submandibular gland: a morphometric study. J Oral Pathol Med. 2014 Aug 26;

[124] Ioanna D. Daskala CCT. Research Article Morphological Changes of Parotid Gland in Experimental Hyperlipidemia.

[125] Pannunzio E, Amancio OMS, Vitalle MS de S, Souza DN de, Mendes FM, Nicolau J. Analysis of the stimulated whole saliva in overweight and obese school children. Rev Assoc Med Bras. 56(1):32–6.

[126] Lamy E, Baptista ES, Coelho AV, Silva FC e. Morphological alterations in salivary glands of mice (Mus musculus) submitted to tannin enriched diets: comparison with sialotrophic effects of sympathetic agonists stimulation. Arq Bras Med Veterinária e Zootec. Escola de Veterinária UFMG; 2010 Aug;62(4):837–44.

[127] Marosti AR, Almeida FN de, Franzói de Moraes SM, Molinari SL, Natali MRM. Effects of the cafeteria diet on the salivary glands of trained and sedentary Wistar rats. Acta Sci Biol Sci. 2012 Jan 3;34(1):113–8.

[128] Takahashi S, Uekita H, Kato T, Yuge F, Ushijima N, Inoue K, et al. Involvement of apoptosis and proliferation of acinar cells in atrophy of rat parotid glands induced by liquid diet. J Mol Histol. 2012 Dec;43(6):761–6.

[129] Takahashi S, Uekita H, Kato T, Yuge F, Ushijima N, Inoue K, et al. Immunohistochemical and ultrastructural investigation of acinar cells in submandibular and sublingual glands of rats fed a liquid diet. Tissue Cell. 2014 Apr;46(2):136–43.

[130] Garrett JR. Effects of Autonomic Nerve Stimulations on Salivary Parenchyma and Protein Secretion. Karger Publishers; 1999;11:59–79.

[131] Cevik Aras H, Ekström J. Pentagastrin-induced protein synthesis in the parotid gland of the anaesthetized rat, and its dependence on CCK-A and-B receptors and nitric oxide generation. Exp Physiol. 2006 Jul;91(4):673–9.

[132] Cevik-Aras H, Godoy T, Ekstrom J. Melatonin-induced protein synthesis in the rat parotid gland. J Physiol Pharmacol. 2011 Feb;62(1):95–9.

[133] Loy F, Diana M, Isola R, Solinas P, Isola M, Conti G, et al. Morphological evidence that pentagastrin regulates secretion in the human parotid gland. J Anat. 2012 May; 220(5):447–53.

[134] De Matteis R, Puxeddu R, Riva A, Cinti S. Intralobular ducts of human major salivary glands contain leptin and its receptor. J Anat. 2002 Nov;201(5):363–70.

[135] Bohlender J, Rauh M, Zenk J, Gröschl M. Differential distribution and expression of leptin and the functional leptin receptor in major salivary glands of humans. J Endocrinol. 2003 Aug;178(2):217–23.

[136] Vincenti DC, Murray GI. The proteomics of formalin-fixed wax-embedded tissue. Clin Biochem. 2013 Apr;46(6):546–51.

[137] De Wijk RA, Prinz JF, Engelen L, Weenen H. The role of alpha-amylase in the perception of oral texture and flavour in custards. Physiol Behav. 2004 Oct 30;83(1):81–91.

[138] Morris C. Impact of calcium on salivary α-amylase activity, starch paste apparent viscosity and thickness perception. Chemosensory Perception. Springer; 2011.

[139] Lamy E. Avaliação das alterações do perfil proteico salivar em indivíduos com obesidade mórbida nas fases pré-e pós cirurgia bariátrica. 2014. p. 124.

[140] Bartoshuk LM, Duffy VB, Hayes JE, Moskowitz HR, Snyder DJ. Psychophysics of sweet and fat perception in obesity: problems, solutions and new perspectives. Philos Trans R Soc Lond B Biol Sci. 2006 Jul 29;361(1471):1137–48.

[141] Marchetti L, Gabrielli MG, Materazzi G, Menghi G. Cellular compartmentation of lysozyme and alpha-amylase in the mouse salivary glands. Immunogold approaches at light and electron microscopy level. Histol Histopathol. 2000 Apr;15(2):337–46.

[142] Ding Q, Zhang Y, Cong X, Cai Z, Han J, Su Y, et al. Confocal microscopy with double immunofluorescence staining reveals the functional transient receptor potential vanilloid subtype 1 expressed in myoepithelial cells of human submandibular glands. Microsc Res Tech. 2012 May;75(5):555–60.

[143] Parkkila S, Kaunisto K, Rajaniemi L, Kumpulainen T, Jokinen K, Rajaniemi H. Immunohistochemical localization of carbonic anhydrase isoenzymes VI, II, and I in human parotid and submandibular glands. J Histochem Cytochem. 1990 Jul 1;38(7):941–7.

[144] Ryberg AT, Warfvinge G, Axelsson L, Soukup O, Götrick B, Tobin G. Expression of muscarinic receptor subtypes in salivary glands of rats, sheep and man. Arch Oral Biol. 2008 Jan;53(1):66–74.

[145] Wang D, Yuan Z, Inoue N, Cho G, Shono M, Ishikawa Y. Abnormal subcellular localization of AQP5 and downregulated AQP5 protein in parotid glands of streptozotocin-induced diabetic rats. Biochim Biophys Acta. 2011 May;1810(5):543–54.

[146] Susa T, Sawai N, Aoki T, Iizuka-Kogo A, Kogo H, Negishi A, et al. Effects of repeated administration of pilocarpine and isoproterenol on aquaporin-5 expression in rat salivary glands. Acta Histochem Cytochem. 2013 Dec 28;46(6):187–97.

[147] Nashida T, Yoshie S, Imai A, Shimomura H. Presence of cytoskeleton proteins in parotid glands and their roles during secretion. Arch Oral Biol. 2004 Dec;49(12):975–82.

[148] Sramkova M, Porat-Shliom N, Masedunkas A, Wigand T, Amornphimoltham P, Weigert R. (2012). Salivary Glands: A Powerful Experimental System to Study Cell Biology in Live Animals by Intravital Microscopy, Current Frontiers and Perspectives in Cell Biology, StevoNajman (Ed.), ISBN: 978-953-51-0544-2, InTech, 2012. Available from: http://www.intechopen.com/books/current-frontiers-andperspectives-in-cell-biology/novel-imaging-approaches-to-study-cell-biology-in-live-animals

[149] Gorr S-U, Venkatesh SG, Darling DS. Parotid secretory granules: crossroads of secretory pathways and protein storage. J Dent Res. 2005 Jun;84(6):500–9.

[150] Elewa YHA, Ichii O, Otsuka S, Hashimoto Y, Kon Y. Structural changes of goat parotid salivary gland: pre-and post-weaning periods. Anat Histol Embryol. 2014 Aug; 43(4):265–72.

[151] Stolte M, Ito S. A comparative ultrastructural study of the parotid gland acinar cells of nine wild ruminant species (mammalia, artiodactyla). Eur J Morphol. 1996 Jan; 34(2):79–85.

[152] Mednieks MI, Szczepanski A, Clark B, Hand AR. Protein expression in salivary glands of rats with streptozotocin diabetes. Int J Exp Pathol. 2009 Aug;90(4):412–22.

[153] Takano K, Malamud D, Bennick A, Oppenheim F, Hand AR. Localization of salivary proteins in granules of human parotid and submandibular acinar cells. Crit Rev Oral Biol Med. 1993 Jan;4(3-4):399–405.

[154] Isola M, Cabras T, Inzitari R, Lantini MS, Proto E, Cossu M, et al. Electron microscopic detection of statherin in secretory granules of human major salivary glands. J Anat. 2008 May;212(5):664–8.